Probability and Mathematical Statistics (Continued)
PURI, VILAPLANA, and WERTZ • New Pe[...] Applied Statistics
RANDLES and WOLFE • Introduction to t[...] Statistics
RAO • Linear Statistical Inference and Its A[...]
RAO • Real and Stochastic Analysis
RAO and SEDRANSK • W.G. Cochran's Impact on Statistics
RAO • Asymptotic Theory of Statistical Inference
ROHATGI • An Introduction to Probability Theory and Mathematical Statistics
ROHATGI • Statistical Inference
ROSS • Stochastic Processes
RUBINSTEIN • Simulation and The Monte Carlo Method
SCHEFFE • The Analysis of Variance
SEBER • Linear Regression Analysis
SEBER • Multivariate Observations
SEN • Sequential Nonparametrics: Invariance Principles and Statistical Inference
SERFLING • Approximation Theorems of Mathematical Statistics
SHORACK and WELLNER • Empirical Processes with Applications to Statistics
TJUR • Probability Based on Radon Measures

Applied Probability and Statistics
ABRAHAM and LEDOLTER • Statistical Methods for Forecasting
AGRESTI • Analysis of Ordinal Categorical Data
AICKIN • Linear Statistical Analysis of Discrete Data
ANDERSON, AUQUIER, HAUCK, OAKES, VANDAELE, and WEISBERG • Statistical Methods for Comparative Studies
ARTHANARI and DODGE • Mathematical Programming in Statistics
BAILEY • The Elements of Stochastic Processes with Applications to the Natural Sciences
BAILEY • Mathematics, Statistics and Systems for Health
BARNETT • Interpreting Multivariate Data
BARNETT and LEWIS • Outliers in Statistical Data, *Second Edition*
BARTHOLOMEW • Stochastic Models for Social Processes, *Third Edition*
BARTHOLOMEW and FORBES • Statistical Techniques for Manpower Planning
BECK and ARNOLD • Parameter Estimation in Engineering and Science
BELSLEY, KUH, and WELSCH • Regression Diagnostics: Identifying Influential Data and Sources of Collinearity
BHAT • Elements of Applied Stochastic Processes, *Second Edition*
BLOOMFIELD • Fourier Analysis of Time Series: An Introduction
BOX • R. A. Fisher, The Life of a Scientist
BOX and DRAPER • Empirical Model-Building and Response Surfaces
BOX and DRAPER • Evolutionary Operation: A Statistical Method for Process Improvement
BOX, HUNTER, and HUNTER • Statistics for Experimenters: An Introduction to Design, Data Analysis, and Model Building
BROWN and HOLLANDER • Statistics: A Biomedical Introduction
BUNKE and BUNKE • Statistical Inference in Linear Models, Volume I
CHAMBERS • Computational Methods for Data Analysis
CHATTERJEE and PRICE • Regression Analysis by Example
CHOW • Econometric Analysis by Control Methods
CLARKE and DISNEY • Probability and Random Processes: A First Course with Applications, *Second Edition*
COCHRAN • Sampling Techniques, *Third Edition*
COCHRA[...]
CONOV[...]
CONOV[...] atistics
CORNE[...] The Analysis of Mixt[...]

Applied Probability and Statistics (Continued)

COX • Planning of Experiments
COX • A Handbook of Introductory Statistical Methods
DANIEL • Biostatistics: A Foundation for Analysis in the Health Sciences, *Third Edition*
DANIEL • Applications of Statistics to Industrial Experimentation
DANIEL and WOOD • Fitting Equations to Data: Computer Analysis of Multifactor Data, *Second Edition*
DAVID • Order Statistics, *Second Edition*
DAVISON • Multidimensional Scaling
DEGROOT, FIENBERG and KADANE • Statistics and the Law
DEMING • Sample Design in Business Research
DILLON and GOLDSTEIN • Multivariate Analysis: Methods and Applications
DODGE • Analysis of Experiments with Missing Data
DODGE and ROMIG • Sampling Inspection Tables, *Second Edition*
DOWDY and WEARDEN • Statistics for Research
DRAPER and SMITH • Applied Regression Analysis, *Second Edition*
DUNN • Basic Statistics: A Primer for the Biomedical Sciences, *Second Edition*
DUNN and CLARK • Applied Statistics: Analysis of Variance and Regression
ELANDT-JOHNSON and JOHNSON • Survival Models and Data Analysis
FLEISS • Statistical Methods for Rates and Proportions, *Second Edition*
FLEISS • The Design and Analysis of Clinical Experiments
FOX • Linear Statistical Models and Related Methods
FRANKEN, KÖNIG, ARNDT, and SCHMIDT • Queues and Point Processes
GALLANT • Nonlinear Statistical Models
GIBBONS, OLKIN, and SOBEL • Selecting and Ordering Populations: A New Statistical Methodology
GNANADESIKAN • Methods for Statistical Data Analysis of Multivariate Observations
GREENBERG and WEBSTER • Advanced Econometrics: A Bridge to the Literature
GROSS and HARRIS • Fundamentals of Queueing Theory, *Second Edition*
GUPTA and PANCHAPAKESAN • Multiple Decision Procedures: Theory and Methodology of Selecting and Ranking Populations
GUTTMAN, WILKS, and HUNTER • Introductory Engineering Statistics, *Third Edition*
HAHN and SHAPIRO • Statistical Models in Engineering
HALD • Statistical Tables and Formulas
HALD • Statistical Theory with Engineering Applications
HAND • Discrimination and Classification
HOAGLIN, MOSTELLER and TUKEY • Exploring Data Tables, Trends and Shapes
HOAGLIN, MOSTELLER, and TUKEY • Understanding Robust and Exploratory Data Analysis
HOEL • Elementary Statistics, *Fourth Edition*
HOEL and JESSEN • Basic Statistics for Business and Economics, *Third Edition*
HOGG and KLUGMAN • Loss Distributions
HOLLANDER and WOLFE • Nonparametric Statistical Methods
IMAN and CONOVER • Modern Business Statistics
JAGERS • Branching Processes with Biological Applications
JESSEN • Statistical Survey Techniques
JOHNSON • Multivariate Statistical Simulation: A Guide to Selecting and Generating Continuous Multivariate Distributions
JOHNSON and KOTZ • Distributions in Statistics
      Discrete Distributions
      Continuous Univariate Distributions—1
      Continuous Univariate Distributions—2
      Continuous Multivariate Distributions

(continued on back)

A DIARY ON INFORMATION THEORY

A DIARY ON INFORMATION THEORY

ALFRÉD RÉNYI

JOHN WILEY & SONS
Chichester · New York · Brisbane · Toronto · Singapore

This work is the English version of the Hungarian
Napló az információelméletről, Gondolat, Budapest

Translated by Zsuzsanna Makkai-Bencsáth
Translation editors: Marietta and Tom Morry of Ottawa, Canada

First published in English by Akadémiai Kiadó, Budapest 1984, Reprinted 1987
Reprint is a joint edition of Akadémiai Kiadó and
  John Wiley & Sons
Copyright © 1984 by Akadémiai Kiadó, Budapest

All rights reserved.

No part of this book may be reproduced by any means, or transmitted, or translated into a machine language without the written permission of the Publisher.

Library of Congress Cataloging in Publication Data:

Rényi, Alfréd.
  A diary on information theory.
  Translation of Napló az információelméletről.
  Bibliography: p.
  Includes index.
  1. Mathematics. 2. Probabilities. 3. Information
theory. I. Title.
QA7.R4613 1987   001.53'9   86—9196
ISBN 0 471 90 971 8

British Library Cataloguing in Publication Data:

Rényi, Alfréd
  A diary on information theory. — (Wiley series on probability and mathematical statistics—applied)
  1. Information theory
  I. Title II. Napló az információelméletről.
  English
  001.53'9   Q360
  ISBN 0 471 90 971 8

Printed in Hungary

Contents

Foreword (Pál Révész)	vii
On the mathematical notion of information	1
Preface	1
The diary	4
First lecture	4
Second lecture	14
Third lecture	23
Fourth lecture	34
Fifth lecture	43
Preparation for my talk	49
Games of chance and probability theory	56
Introduction	56
On the shuffling of cards	56
Problems on the distribution of cards	60
Game strategies	66
A mathematician's war against the Casinos	74
Bibliography	75
Notes on the teaching of probability theory	77
1. Why should probability theory be taught?	77
2. What should be taught?	79
3. How should probability theory be taught?	81
Variations on a theme by Fibonacci	85
The theme	85
Final cadence	104
The mathematical theory of trees	105
Introduction	105
Enumerating problems relating to trees	107

CONTENTS

Statistical theory of trees ... 111
Applications in operation research ... 113
Trees and information theory ... 115
Trees and permutations ... 118
Trees and chemistry ... 119
Trees and biology ... 120
Bibliography ... 121

Index ... 123

Foreword

At one point in *Dialogues on Mathematics* by Alfréd Rényi Archimedes is explaining to King Hieron that only those who like mathematics for itself can apply its results successfully, and he says: "... mathematics rewards only those who are interested in it not only for its rewards but also for itself. Mathematics is like your daughter, Helena, who suspects every time a suitor appears that he is not really in love with her, but is only interested in her because he wants to be the king's son-in-law. She wants a husband who loves her for her own beauty, wit and charm, and not for the wealth and power he can get by marrying her. Similarly, mathematics reveals its secrets only to those who approach it with pure love, for its own beauty. Of course, those who do this are also rewarded with results of practical importance. But if somebody asks at each step, 'What can I get out of this?' he will not get far. You remember I told you that the Romans would never be really successful in applying mathematics. Well, now you can see why: the yare too practical." (From A. Rényi: *Dialogues on Mathematics,* Holden-Day.)

The entire oeuvre of Rényi is permeated with his love of mathematics. Throughout his life he was devoted to mathematics as a lover is to his beloved. But what truly characterizes Rényi and makes him unique among outstanding mathematicians is not this unconditional love of mathematics, which is common to all true mathematicians. What distinguishes Rényi as a mathematician was that he loved people with the same fervour with which he loved mathematics so that he wanted to make mathematics a gift to them, a source of pleasure and joy.

Devoted to his research work, he was always striving at the same time to bring mathematics close to as many people as possible. Sometimes he accomplished this by solving concrete practical problems, thus demonstrating that modern mathematics can have many applications. He was the first in Hungary to construct a mathematical model in economics, and to achieve results in the fields of chemistry and biology by using the theory of stochastic processes.

On other occasions, he tried to bring mathematics closer to the layman in writings which even from a literary point of view are excellent. Those which are probably best known are the two books entitled: *Dialogues on Mathematics* and *Letters on Probability*.

The first part of the present book, entitled: "On the Mathematical Notion of Information", is intended to be a continuation of those two books. Unfortunately, it was left unfinished at his untimely death. The final chapter of the text given here was completed from Rényi's own notes by one of his pupils, Gyula Katona. I think the reader will find this book a natural continuation of the previous two. This one, too, was not written with the aim of teaching a particular field of mathematics. Its intent is to explain what mathematics is, what it can contribute to our everyday lives, how it can further the development of the way we think and how we can enjoy its beauty.

There are striking similarities among these works in style as well. For example, in this most recent work, it is again not the author himself who speaks. Instead, he uses another person, this time an imaginary student of his, to convey his thoughts. Rényi was very proud of the fact that many readers of his *Letters on Probability* believed Pascal to be the author of the letters. For example, he used to tell with pleasure how, when a French publisher was having the book translated, the translator, after reading part of it, asked why it needed to be translated at all since Pascal must have written the original in French.

I am sure that Rényi would also be pleased to hear that some reader of this current book had believed himself to be reading the actual diary of a university student.

Although I am sure that Rényi's style makes it quite possible to believe that one has before one the writings of a young man who was just becoming acquainted with his subject, readers, if they do not let themselves be led astray by the style, but recognize the depth of thought being presented, must realize that this "diary" could only have been written by one who understands the whole subject in all of its depth as well as the breadth of its technical complexities.

I would like to highlight one more similarity which connects these three works by Rényi. When these books were being written there were many, sometimes very heated, disputes taking place among Hungarian mathematicians about the application and teaching of mathematics, the relation between pure and applied mathematics and many other questions every mathematician is interested in.

Rényi thought it necessary to communicate his ideas on these questions not only to a limited group of mathematicians but also to explain these problems and his opinions about them to anyone who might be interested. I would therefore ask the reader to remember that he will not meet only widely accepted truths in this book but also opinions the author fought very hard to have accepted. He wanted this work to be a weapon in that battle.

This volume also contains some other popular articles by Rényi, most of which appeared in various journals. Their message too is how one can come to like mathematics. One needs some mathematical knowledge to be able to read some of them, yet, even if some readers lack the necessary knowledge and cannot follow them all the way through, they will certainly find sufficient incisive thoughts in them to make them enjoyable. (This remark relates to the diary too.) For example, at the end of "Games of Chance and Probability Theory", several mathematical tools are employed which are not generally known as they are not taught in high school or at university. Nevertheless, I feel that all readers can benefit from the first part of the paper. An attempt has been made to explain the technical details in a series of footnotes written by Gyula Katona. It is hoped that they will be of assistance to the reader.

Pál Révész

On the mathematical notion of information
Diary of a university student

PREFACE

A couple of years ago, just after the oral examinations, I found in my mail a heavy bundle of papers accompanied by a letter from one of my recently graduated students. The letter read:

"Dear Professor, I am sending this draft to you to fulfill the promise I made during our talk after my oral examination. Let me recall that conversation for you since I remember it very clearly. While it was very important for me because of its subject, I would not be at all surprised if you have forgotten its details since you are busy with many other things.

The title of my thesis was: "The mathematical notion of information". After I had successfully defended it you encouraged me to publish it because, as you explained, no good book or monograph has as yet been published in Hungarian discussing the fundamental problems of information theory at an adequate level while still being understandable to the layman. For this purpose, you continued, I should change the style of the thesis, which is now too theoretical and dull for a non-mathematician. Moreover, it assumes a background knowledge that most laymen do not possess. I replied that I would be more than glad to do so. I even indicated that I had based the thesis on a diary which I had kept during my studies of information theory and in which I had recorded all of my thoughts on the subject, later "translating" it into the usual technical thesis-language. And I remarked that while the diary was not ready for printing in its original form, its tone was much more appropriate for the purpose so that it would be more profitable for me to work on that than the thesis. You answered that you would be glad to look into this diary, if there were no personal remarks I didn't want anyone to see. So I promised to send it to you.

Since then, I have been notified that I got the job at the Univ. of Madagascar that UNESCO had been advertising. When you receive my letter, I will be on my way to Madagascar, probably for three years. The main reason I applied for this job was that after my exam I got married and we don't have an apartment. So now this problem will be taken care of for three years and I even hope to be able to save enough money while we are there to buy an apartment when we return. This place may not be the most appropriate for the scientific work which you encouraged me to pursue, but I hope that I'll have time to work on the problems you mentioned and I'll report the results, if any, to you. I'd greatly appreciate it if you'd be so kind as to take the time to answer my letters and give me your suggestions on my work.

As far as the enclosed diary is concerned, I doubt that I'll have the time in Madagascar to work on it. Therefore I would like to ask you to look into it and see if it can be used in any way. If you think so, would you please ask someone else to make the necessary changes. I would be glad if my diary, changed or not, could be published, but I would ask you by all means to keep my name a secret. My reasons for this request are the following: 1. I have no illusions about the originality of the thoughts written in the diary. They were stimulated by the lectures I attended while I was at the University and therefore they are probably somewhat more the intellectual property of my professors than mine. 2. Those which I feel are my own original thoughts are the ones which relate to education at the University. I don't know if you will leave those in or cross them out. In the former case, it is even more important that my name not be connected with the diary, since I hope to get a position at the University on my return from Madagascar. I'm afraid it would not be wise to have my critical remarks published under my own name. (Perhaps my anxiety seems excessive but unfortunately, I have learned the hard way that it's better to be cautious.)

I would like to thank you again for your help and attention during my studies,

Sincerely yours:
Bonifác Donát".

After reading the diary, I was convinced that it should be published without any modification, rewriting or deletion. Any change would only decrease its authenticity and freshness. Of course, I disagree with Donát's rationale for keeping his authorship anonymous. First of all, it is not true that his diary reflects only ideas he had heard at the lectures: there are many really original thoughts presented in novel ways which should be considered his own. As for the critical remarks, even if they are not mature they should be considered seriously, because their aim is the improvement of university education. Since education is for the students and not vice versa, students do have the right to form and express independent opinions about education, the way they are taught. And if they take the time to think about it seriously, no one should reproach them for doing so, even if he doesn't agree with them. Therefore

I find Donát's concern, that there might be problems with his future university position because of these critical remarks, unfounded. I can surmise, of course, what kind of bad experiences Donát was referring to. It so happens that I wanted to hire him as a teaching assistant. For this reason several members of the Department interviewed him and as is usual in such cases, asked him about his studies, extracurricular activities, etc. In the end, unfortunately, nothing came of the appointment for budgetary reasons, but it seems Donát attributed this to other reasons.

However, even though I disagree with his reasons, I will obey his wish. Accordingly, I have given him the pseudonym of Bonifác Donát[1]. Of course, if he is willing to shed his anonymity in a few years, I will be glad to attest to his authorship. Until such time, let him remain anonymous, but his diary should be published[2]. I hope that it will help many thoughtful people to find their way through the difficult but important fundamental questions of information theory, and, as Donát did, formulate their own views based on the facts and arguments set out therein.

Alfréd Rényi

---

[1] Those who know Latin will realize that this name refers to the diary given to me by my student.

[2] I will publish only the first part of the diary, the one which deals with fundamental notions of information theory. The second part concerns information theoretical theorems using a more extended mathematical apparatus and so would not be understandable to the layman. This part, in any case, is in quite a different style. The "diary-like" character gives way to something like that of a causal notebook. Probably because Donát was by then so involved with what for him were the new, strictly mathematical problems and their accompanying difficulties, that he did not have the time to reflect on the more fundamental questions. This is quite unfortunate, because such reflections are exactly what makes the first part so interesting and instructive for both kinds of readers, mathematician and non-mathematician alike.

THE DIARY

> I keep six honest serving-men
> (They taught me all I knew);
> Their names are What and Why and When
> And How and Where and Who.
>
> (Kipling)

First lecture

Today was the first lecture on information theory. The lecturer seems good. He started by saying that this field of mathematics is not even twenty years old*, i.e. when we (who were sitting there) had been born information theory didn't exist. Probably it's very impulsive of me to take such a fancy to an idea, but at that moment, before I had heard anything further about information theory, I decided to study it very seriously. Right away I decided to keep a diary instead of ordinary lecture notes on this subject. That is, I will put into writing not only what I hear in the lectures but my own thoughts as well, both the questions I ask myself, and, of course, the answers, too, if I can find any. That is why I chose Kipling's poem as a motto. As I listened to the lecturer, this intention grew stronger in me. I liked it very much when he said that he was going to organize the lectures as seminars, that he wanted us to participate in them actively, by asking questions whenever something is not clear and that he would question us, from time to time, too. I like this much better than those courses where the lecturer just pontificates and our only role is to listen in silence. When a lecturer lets anyone interrupt his talk, that means he is not afraid of what we might ask him, something he might not be able to answer. Of course, it is quite stupid if a lecturer is really afraid of this, because we don't expect a teacher to know everything. Someone who knows his subject doesn't get embarrassed in such situations. Last year, for example, one of my classmates asked a question of a professor who said it was a new and very interesting problem that nobody had solved as far as he knew. He promised to think about it some more. At the next lecture, he said that he had found a solution, which really had not been known before and that he was going to include the problem and its solution in his latest book which was soon to be published with a note that it was my classmate who had formulated it. Unfortunately, there are professors who react quite differently. In our second year, we had one who once gave a false proof for a theorem.

---

* The manuscript was written in 1969. (Publisher)

When a few of us pointed out that we didn't understand a certain step in the proof (the one which was wrong), he told us offhandedly to pay more attention next time.

The lecturer started by speaking about the importance of the notion of information. He said he was sure we were adequately informed about this matter so that he would only talk about it briefly. As he pointed out, information is transmitted in every living system, like the human body; the sensory systems collect information from the outside world to be transmitted via the nervous system to the brain. The brain, after some information processing, sends commands, which are information too, to the muscles via nerve cells again. Similarly, in a factory or any other organization where many people interact, there is a continuous flow of information in the form of reports, orders, queries without which collaboration is not possible. He remarked that in every great achievement of modern technology, the transmission, processing, and storage of information play the main role. For example, one of the principal problems in space flight involves the transmission of information between the spacecraft and the control center. The essence of the computer is that it can process a large amount of information at great speed in accordance with a given program (which is itself a form of information). One of the chief problems of automation is that of the information exchange among the different parts of the machine. Feedback, for example, means that the central processing unit receives information about how the machine is carrying out its commands and, based on this information, modifies its instructions appropriately.

The lecturer indicated that the mathematical theory of information had come into being when it was realized that the flow of information can be expressed numerically in the same way as distance, time, mass, temperature, etc.

He explained this by referring to the game known as "Bar-kochba"*.

One can measure the information needed to guess the message that the others have decided on by the number of questions required to get that information when one is using the most effective system of questioning. "Question", according to the rules of the game, means a question that can be answered with a "yes" or a "no". If we write the answers down by writing 1 for a "yes" and 0 for a "no", the answer-series (which characterizes uniquely the "something or somebody" we have to guess) is replaced by a sign sequence. This procedure is called *coding* and the sign sequences of 0's and 1's are called

---

* (Translator's note: this game is similar to "Twenty Questions". See p. 13.)

codewords. It is well known that every natural number can be expressed with ones and zeroes, i.e., written in the binary system. Briefly:

> 1, 2, 3, 4, 5, 6, 7, 8 in the binary number system take the form of: 1, 0, 11, 100, 101, 110, 111, 1000, i.e., when adding 1, the final digit has to be increased by one; but if the result would be two, the digit should be "changed" to zero and a 1 written in the next left position, etc. So going from right to left, a 1 in the first position means 1, in the second it means 2, in the third, $2^2=4$, in the fourth, $2^3=8$, etc. For example, 1110 is the binary expression of $8+4+2=14$.
>
> (Gy. Katona)

But it is evident, too, that any text can be expressed or coded in a sequence of 1's and 0's. This can be done, for example, by associating a 0, 1-sequence with every letter in the alphabet and then by rewriting the text letter by letter. If we start with the alphabet a b c d e f g h i j k l m n o p q r s t u v w x y z of 24 letters and include the period, space, semicolon, comma, colon and quotation mark, then we need 32 such sequences. There are exactly 32 possible sequences made up of 5 digits of 0's or 1's; therefore we can assign one such 5-digit sequence to every letter and symbol. Now the text will consist of five times as many signs as before. Every message, every piece of information may thus be encoded in a sequence of zeros and ones. The practical consequence of this fact lies in computer programming. The "mother tongue" of a computer is the binary number system and therefore not only the data but all instructions have to be coded in sequences of 1's and 0's. It seems reasonable to measure the amount of information in a message by the number of signs needed to express it in zeros and ones (using the most appropriate way to code, i.e., to get the shortest sequence). The lecturer emphasized that this is not yet the precise mathematical definition that we would see later on. It is only a way of expressing what is meant by an amount of information, and it can help us to get an idea of this concept. We should consider what he had said to be a first step towards the concept of information. He stressed that when we want to measure information, to quantify it with numbers, we deliberately ignore features like its content and importance.

The example he gave to clarify this was a great successs (especially with the girls): the answer to the question, "Miss, do you like cheese?" – regardless of whether it is a yes or a no – consists of 1 unit of information, but so does the answer to, "Miss, will you marry me?" although the importance and the content of the two replies are very different. He pointed out that according to what he had said so far, the answer to a question which can only be answered with a "yes" or "no" contains one unit of information, the meaning of the particular question being irrelevant.

To put this another way: the unit of information is the amount of information which can be expressed (coded) with one 0 or 1 only. Accordingly, if we write an arbitrary number in the binary system, the measure of information of every digit is 1. This is the way that the unit of information got its name: it is called a "bit", the abbreviation of "binary digit" (i.e., a digit in a number expressed in binary form). There is also a little pun hidden in this name, since "bit" means a small piece or morsel. So a "bit" is a morsel of information.

In general, he told us further, coding is necessary for the transmission of information. The method of encoding depends on the type of transmission. For example, for the transmission of telegrams, the message is coded in the Morse code, in which sequences of dots and dashes are assigned to letters. The TV transmitter codes the picture by decomposing it into small dots. According to how dark or light a particular dot is it emits an appropriate signal which is then transmitted via electromagnetic waves to the TV receiver to be transformed (i.e. decoded) into a picture. One can also give a much simpler example of coding, namely writing, where letters are assigned to phonemes and reading the text corresponds to decoding it. Information is coded in our brains, too, before it is stored in our memory, although we don't know yet how this coding/decoding is accomplished. Putting a knot in your handkerchief to help you remember some specific thing is also coding. The lecturer mentioned that some Indian tribes managed to develop this kind of coding to a very sophisticated level, so that they could send messages to each other by means of knots on a string. Again, it is coding when tourists, trapped in the Alps, sends SOS signals with whistles or when ships communicate with light signals. All kinds of secret codes are, of course, examples of coding. In fact, this is the origin of the word "coding" itself. A record or a tape contains speech or music in a coded form. One of us remarked that this just goes to show that we are in the same situation with coding as the character in the play by Moliere, who didn't know that he had been speaking prose all his life. In the same way we didn't know that we are coding and decoding in almost all of our activities.

There were exactly 32 of us attending this lecture. The lecturer asked if he had one of us in mind, how many questions would be needed to find out who that person was. I realized the answer right away: 5 questions. He asked what strategy I would use. I answered that I would write down the names of all persons in the class in alphabetical order and, as a first question, I would ask if the name was among the first 16. Regardless of whether the answer were "yes" or "no", the number of the possible names would now be only 16. Similarly, with the second question the number of remaining possibilities would be reduced to 8, with the third to 4, and after the fourth there could be only two, so that by asking the fifth question I would be able with certainty to determine which of the 32 students he had in mind. Then he asked me what would happen if I had to ask all five questions one after the other, without getting an answer before all the questions were asked. Would 5 questions be enough then? My answer was that I didn't know, because when I played "Bar-kochba", I always decided on the next question after considering the foregoing answers, but that my hunch was it would take more than 5 questions. But I was wrong! The professor explained to us how one can ask 5 questions so that one can figure out from the answers who the one person is. One has to number the students from 0 to 31 and write the binary representation of each of these numbers. There are now 32 sequences, each consisting of 5 digits of 0 or 1, corresponding to the 32 students.

(The numbers expressible by 1, 2, 3, or 4 digits are to be completed by zeros to give five digits, for example, 0=00000, 1=00001.)

I am the 14th on the list, and so the number 01110 corresponds to my name. (Bonifác Donát is referring of course, to his real name. He is not aware of the pseudonym.) So the 5 questions should be as follows: is the first, second, etc., digit in the binary "codename" of that person a 1? If the answers, in order, are "no, yes, yes, yes, no", then I was the one whom the professor had in mind. I was really surprised, since I had thought of myself as an expert in "Bar-kochba". We used to play it a lot in the dormitory and I was considered to be the best. (Recently I made quite an impression by figuring out "the hole dug by a worm in the apple which fell onto Newton's head".) Yet I would never have thought of asking the questions rapidly one after the other, without waiting for the answers.

Then our lecturer cited some more examples, such as, "how many bits of information are there in a passport number?" There are about 7.5 million adult citizens in Hungary with a passport. So the question was, if he should think of one of them, how many questions would be needed to figure out who that person was? We answered quite promptly. Since $2^{22}=4{,}194{,}304$ and $2^{23}=8{,}388{,}608$, the number of questions should be 23. Then he asked if it would be correct to say that a passport number has exactly 23 bits of information. Some of us said that would be true only when the number of citizens reaches 8,388,608, whose base 2 logarithm is exactly 23. If there are only 7 and a half million passports right now, then the information in a passport number is a little bit less than 23 bits but certainly more than 22 bits. And together we concluded that it we modify the rules of "Bar-kochba" so that one can think of only one of exactly N different things (such as a person, or an object etc.) then the information needed to figure the message out is $\log_2 N$.

Now he suggested that we try to phrase this result without any reference to the game. After some experimentation, we ended up with the following: *Let us consider an unknown x of which we know nothing except that it belongs to the set H having N elements. This information amounts to* $\log_2 N$ *bit.* This is called *Hartley's formula*. He followed this by explaining the *law of the additivity of information*. This is again easier to understand through reference to "Bar-kochba". Assume that we have to guess two things, let us say x_1 and x_2, knowing only that x_1 belongs to the set H_1 of N_1 elements and that x_2 belongs to the set H_2 of N_2 elements. We can say that we have to guess the pair (x_1, x_2) which belongs to set H of all possible pairs (x_1, x_2), where x_1 is an arbitrary element of H_1, and x_2 is an arbitrary element of H_2 independently of x_1. Obviously, the set H has $N_1 \cdot N_2$ elements. Therefore the number of questions needed to guess the pair (x_1, x_2) (in other words the amount of information needed to determine (x_1, x_2)), according to Hartley's formula is $\log_2 (N_1 \cdot N_2)$. On the other hand, we can guess x_1 and x_2 separately. We need $\log_2 N_1$ questions to figure out x_1 and $\log_2 N_2$ for x_2. Altogether, $\log_2 N_1 + \log_2 N_2$ questions will be necessary to guess x_1 and x_2, which means that the necessary amount of information to find x_1 and x_2 is $(\log_2 N_1 + \log_2 N_2)$ bits. It looks as though we have two formulas for this information, but the two are equivalent (because of the well-known property of logarithm functions, that the logarithm of a product is equal to the sum of the logarithms of the factors of the product):

$$\log_2 N_1 \cdot N_2 = \log_2 N_1 + \log_2 N_2.$$

This is the law of additivity of information.

At the end of the lecture, we talked a little more about the concept of information. The professor warned us that since we can measure the quantity of information numerically, we can use the word "information" in two ways — one concrete, the other abstract, i.e., one denoting quality, the other quantity. By information we mean the concrete message (information) on the one hand and, on the other hand, its numerical value, i.e., the measure of the abstract amount of information contained in the concrete information, measured in bits. It is better to speak of only the concrete information as "information" and to call its numerical information content the "amount of information". To avoid any misunderstanding, he said we would sometimes use the term "message" instead of "information". It is also worthwhile to consider whether information (in the qualitative sense) has only one meaning for us. He asked us to write down on a piece of paper 10 words whose meaning is close to the meaning of "information". I wrote down the following ten:

1. notice
2. message
3. communication
4. news
5. data
6. instruction
7. enlightenment
8. knowledge
9. characterization
10. announcement.

The others gave some more variants of these words: telecommunication, communique, reference service, data transmission, notification, description, declaration, proclamation, statement.

We had to acknowledge that it would be pretty hard to define the concept of "information" formally, and that it was not necessary to obtain any cases because we all use the term thinking of essentially the same thing and, finally, that we will be defining the precise mathematical concept later on. Our lecturer emphasized again that we had still not reached a definition of this concept but only taken the first step toward it; the same first step taken by Hartley in 1928.

Finally, we solved a few problems to help us understand Hartley's formula better. We considered, for example, the following question (which, as the professor said, belongs to an important chapter of information theory called search theory): we have 27 apparently identical gold coins. One of them is false and lighter (having been made of some other metal and then gold-plated), but it is visually indistinguishable from the others. We also have a balance with two pans, but without weights. Accordingly, any measurement will tell us if the loaded pans weigh the same or, if not, which weighs more. How many measurements are needed to find the false coin? Before solving the problem we were able to set a lower limit to the number of necessary weighings by establishing the following information: the false coin can be any of the 27 coins, therefore the information needed to find it, according to Hartley's formula, is $\log_2 27$.

Since every measurement can have one of three possible outcomes (the left side is heavier, the right side is heavier, the weights of the two are equal), each can give at most $\log_2 3$ information. If we complete x measurements, $x \log_2 3 \geqq \log_2 27$ has to be satisfied. Since $\log_2 27 = 3 \log_2 3$ it follows that $x \geqq 3$, which means that at least 3 measurements are necessary. And we can certainly find the false coin with 3 measurements. First we put 9 coins into each pan. If one side rises, the false coin is among the 9 coins on that side; if the sides balance, then it is among the remaining 9 coins. Performing one measurement has decreased the possibilities to 9. Now choose two sets of three from that 9 and put them into the pans. The result of this second measurement will leave us with only 3 suspicious coins. Lastly, we put one of these 3 on each side of the balance and whatever the result of the weighing, we will know which is the false coin.

After the lecture, I had the feeling that I had understood everything I had heard. Now that I have gone through my notes, I see a lot of problems.

It is obvious that the answer to a question which can be answered with a "yes" or a "no" contains one unit, i.e., one bit of information. But what happens in Bar-kochba when I ask poor questions and after the fifth cannot figure out something that should only require five. How could this happen? The answers to the five questions should have supplied me with 5 bits of information in total. Here I see an apparent contradiction. If 5 bits is enough to guess one of the 32 elements of a set, why is it that sometimes this information is not in fact enough? The amount of information cannot depend on how well I question. In other words: if I ask clumsily, I get less than 5 bits of information from the five answers — let's say 3 bits. But the answers contain one bit each, so where then did those 2 bits disappear to? This question looked very mysterious at first but then I remembered what had happened the last time we played Bar-kochba and that helped me to solve this riddle. What had happened was that that night I was tired and preoccupied, and that was probably why, during the questioning, I asked a question which had been answered a minute before. "But you already asked that", the other told me and suggested that I stay quiet if I was tired — and that was exactly what happened. I was a bit irritated at the time, but now this same event came to my assistance. I realized that the simplest way to question "unwisely" is to repeat a previous question. In such a case, of course, it is still true that both answers, the one to the first question and the one to the second (identical to the first) contain one bit of information, but the second bit is not new information.

So the two answers to the unskilled player's questions contain not two but only one bit of information altogether.

Similarly, when one doesn't exactly repeat a question, but asks inexact ones, one will not get the same information twice, yet only a part of the one bit of information obtained from the second answer will be new. The rest will already have been contained in the answer to the first question. This means that the 2 bits overlap. For example, if half of the 2 bits overlap, what we can get in total from the two answers is not 2, but only $1^1/_2$ bits of information. Assume that we have to guess one of the first 8 numbers. As a start, I ask if the number is one of 1, 2, 3, or 4. Then, whatever the answer, my second question is whether the number is among 1, 5, 6 and 7. The second answer will no doubt give new information but it will be less than one whole bit. It is accordingly possible because of this inappropriate question, that I will need four questions instead of the minimum three required. But if I try with the second question to locate the number among 1, 2, 5 and 6, then I will surely figure it out with the third.

Moreover, it is possible for me to guess the correct answer after hearing the answer to the second question using the incorrect method, while this would be impossible with the correct method of questioning, where I always need the third question! There is a haphazard element in the inexact method. My feeling is that, on the average, one loses using such a system, but I want to think about that some more later on. For the moment, I want to think about another problem, because I think I have to get it straight in my mind to understand the previous one clearly. What does a non-integer amount of information mean? The statement: to guess an arbitrary element of a set H having N elements, the information required is $\log_2 N$ bits is clear if $\log_2 N$ is an integer (i.e. $N=2^k$, where k is a positive integer). We can certainly guess the unknown element with exactly k questions, but not less. But what is meant by $\log_2 N$ questions when $\log_2 N$ is not an integer? Somehow the lecturer didn't touch on this point. Perhaps he will turn to it later, but I was bothered by it now.

I thought about this problem for a long time, and then I knew the solution. My logic was this: to guess an element of a 7-element set one needs 3 questions, and for a 9-element set, one needs 4. The Hartley formula states that the information required is $\log_2 7 = 2.80735$ bits and $\log_2 9 = 3.16993$ respectively. But if I want to guess an x_1 element of a 7-element set H_1 (for example, a particular day of the week) and an x_2 element of a 9-element H_2 set simultaneously, then the number of questions needed is not $3+4=7$ but only six, because there are $7 \cdot 9 = 63$ possible x_1, x_2 pairs altogether, of which I have to find one and $63 < 64 = 2^6$ (or $\log_2 63 = \log_2 7 + \log_2 9 = 5.97728 < 6$). Now I was able to give a general answer to the following question: what does it mean, given an arbitrary number N which is not an integer power of 2, that to guess an arbitrary element of an N-element set requires $\log_2 N$ questions (where $\log_2 N$ now is not an integer!)? I reasoned this out as follows. If I need to guess an unknown element of an N-element set not once but many times, let's say k-times (for example, if I play Bar-kochba with k players each of them deciding on one element of the N-element set independently and I have to guess these k elements), then I can ask questions regarding the unknown k-tuple $(x_1, x_2, \ldots, x_k)$ instead of questioning for k independent unknowns. Since, there are N^k different possibilities (i.e., k-tuples), the number of necessary questions can be calculated by taking the 2-based logarithm of N^k, and rounding it to the next integer. If this number of questions is denoted

by S_k then
$$\log_2 N^k \leq S_k \leq \log_2 N^k + 1.$$

Since $\log_2 N^k = k \log_2 N$, we have
$$\log_2 N \leq \frac{S_k}{k} < \log_2 N + \frac{1}{k}.$$

Since S_k is the number of questions necessary to guess a set of k elements, $\frac{S_k}{k}$ is the number of questions needed for guessing one element *on the average*. By choosing a large enough k, $\frac{1}{k}$ can be made as small as we wish.

The above result means that to guess one element of an N-element set (where N is not a power of 2), the number of questions necessary will be (if we play often enough) on the average, arbitrarily close to $\log_2 N$.

In this sense, it is true that the number of questions needed to guess one of a 7-element set is 2.80735. For example, in the case of $N=7$, since $7^6 = 117,649 < 2^{17}$, to guess six elements of this set simultaneously would require 17 questions while for one element, $\frac{17}{6} = 2.833$ will be enough on the average.

I did some research on the Bar-kochba game to see who Bar-kochba was and why the game was named after him. In 135 B.C. the Jews started a war of independence against the Romans under the leadership of Bar Kochba (whose name means "Son of the Star"). The Romans, in superior numbers, laid siege to a fortress which was defended heroically by Bar Kochba at the head of a small garrison.

So far, this is historical fact. It is also said that Bar Kochba sent out to the Roman camp a scout who was captured and tortured, having his tongue cut out. He escaped from captivity and reported back to Bar Kochba, but being unable to talk, he couldn't tell in words what he had seen. Bar Kochba accordingly asked him questions which he could answer by nodding or shaking his head. Thus he acquired from his mute scout the information he needed to defend the fortress.

The only problem with this very persuasive story is that no historical sourcebook mentions it. The legend was probably made up by the person who invented the game but I couldn't trace who that was. It seems as though it came into being in Budapest at the beginning of this century. At any rate, the game was extremely popular in Budapest at that time, mostly among writers. Both Karinthy and Kosztolányi* mention it several times in their writings. They were masters of the game, along with István Szomaházy.

It occurred to me that, if the story of Bar Kochba were true, then he would have been the forefather of information theory. But there is probably no historical foundation to this legend. Still, it would be interesting to find out how long it has been known that all information can be expressed with yes-no answers (can be coded into a sequence of two symbols).

It seems to have been known for a long time, e.g., an old Indian legend points to this fact. Although information theory is a very young science, its antecedents go back far into the past. This is just another example of the thought at the beginning of Thomas Mann's novel "Joseph and his brothers": "Profoundly deep is the well of knowledge...".

Second lecture

The professor actually started this time with the very point I had come to realize: he explained what Hartley's formula expresses when $\log_2 N$ is not an integer. Then he pointed out that up to this point, when we were speaking about guessing an unknown element of an N-element set, we were implicitly assuming all N elements to be equally probable. In reality, this is rarely the case. When we say ξ is an unknown element of the set $H = \{x_1, x_2, ..., x_N\}$, this means that ξ is a random variable with possible values $x_1, x_2, ..., x_N$. Let us denote p_k the probability that ξ assumes the value x_k (for $k = 1, 2, ..., N$). In general, $p_1, p_2, ..., p_N$ are arbitrary positive numbers such that their sum is 1. When we find out which of the possible values ξ takes, or, in other words, when we observe the random variable ξ, this observation contains a certain amount of information $H(\xi)$. To this point we have talked about how this information can be calculated if ξ assumes all its possible values, i.e., $x_1, x_2, ..., x_N$ with equal probability $p_1 = p_2 = ... = p_N = \frac{1}{N}$.

In this special case, Hartley's formula

$$H(\xi) = \log_2 N$$

holds. In the general case, the so-called Shannon formula applies:

$$H(\xi) = p_1 \log \frac{1}{p_1} + p_2 \log \frac{1}{p_2} + ... + p_N \log \frac{1}{p_N}.$$

---

* Famous Hungarian writers in the 20th century.

The meaning of $H(\xi)$ in the general case is essentially the same as in the equiprobable case, with one small modification. In the language of the Bar-kochba game, $H(\xi)$ can be interpreted as follows: if ξ denotes the entity to be guessed, i.e., the x_1 with probability p_1, the x_2 with probability p_2 and so on, up to the x_N with probability p_N of which the other player is thinking, then playing the game often enough and using the most effective method of questioning, on the average, one needs $H(\xi)$ questions, or more precisely, the number of questions needed is arbitrarily close to $H(\xi)$ with a probability arbitrarily close to 1. In contrast to Hartley's special case, we now have to use the qualification "with a probability arbitrarily close to 1", which is not needed in the symmetric case.

Using the concept of coding instead of the 'game-language', $H(\xi)$ can be defined as follows: let ξ be a random variable which assumes the values $x_1, x_2, ..., x_N$ with probabilities $p_1, p_2, ..., p_N$, respectively. If we have independent observations of the value of ξ and code them with 0-1 sequences, then it can be seen that, on the average, the amount of 0's and 1's for the coding of one observation, with a probability arbitrarily close to 1, will be:

$$H(\xi) = p_1 \log \frac{1}{p_1} + p_2 \log \frac{1}{p_2} + ... + p_N \log \frac{1}{p_N}.$$

We verified this statement using the following example: two coins are flipped and ξ denotes the number of heads. ξ can therefore have 3 values, 0, 1 or 2, with probabilities $\frac{1}{4}, \frac{1}{2},$ and $\frac{1}{4}$, respectively. We have to demonstrate that after observing the value of ξ in many experiments, the resulting sequence can be coded with 0's and 1's in such a way that, on the average, the number of symbols required to code a result is arbitrarily close to:

$$\frac{1}{4} \log_2 \frac{1}{\left(\frac{1}{4}\right)} + \frac{1}{2} \log_2 \frac{1}{\left(\frac{1}{2}\right)} + \frac{1}{4} \log_2 \frac{1}{\left(\frac{1}{4}\right)} = 1.5,$$

with a probability arbitrarily close to 1. One way to do this is as follows: if one toss produces a result of one heads, i.e., if $\xi = 1$, we code it with a 1; if $\xi = 0$ (i.e., if neither coin lands heads up), the result is coded as 00, and for $\xi = 2$, we use 01.

This way, the observations can be coded using either one or two digits depending on the value of ξ. Since $\xi = 1$ with probability $\frac{1}{2}$, the length of any code word corresponding to one toss is a random variable η which assumes 1 with probability $\frac{1}{2}$ and 2 with probability $\frac{1}{2}$ and its expected value is $1.5 = \frac{1}{2} \cdot 1 + 1 \cdot \frac{1}{2} \cdot 2 = 1.5$. Therefore, according to the law of large numbers, with a sufficiently large number of experiments the average length of codewords will be less than $1.5 + \varepsilon$ with a probability arbitrarily close to 1, no matter how small the positive number ε is. This coding process translated into 'Bar-kochba language' has the following meaning: if we have to guess how many heads the other player has tossed, first we should ask whether the result was one head. If the answer is 'yes', we have found out the answer with one question. If the answer is 'no', the second question should be whether the number of heads is two. No matter what the answer is, we will now know how many heads there are, because in case of a 'yes', the answer is 2, while for 'no', it is 0. One can now see why this is the best method of questioning. The question asked should, if possible, be one to which a 'yes' or a 'no' answer is equally probable. If there is no such question one must find one where the probabilities are as close to each other as possible. In this way, we can also see that the Shannon formula applies in the general case.

We then proceeded to investigate another example: a coin is tossed until the first time the outcome is the same as one of the previous ones, i.e., heads or tails again.

ξ now denotes a whole sequence of observations. If the first two tosses are the same, then we can stop right away. On the other hand, if the first two are different then we are sure to have the desired situation after the third toss, which must of necessity have a result identical with either the first or second result. ξ therefore can have the values HH, TT, HTH, HTT, THH and THT (where H stands for Head and T for Tail) and the probabilities of these events are $\frac{1}{4}, \frac{1}{4}, \frac{1}{8}, \frac{1}{8}, \frac{1}{8}$ and $\frac{1}{8}$ respectively, so

$$H(\xi) = 2 \cdot \frac{1}{4} \log_2 4 + 4 \frac{1}{8} \log_2 8 = 2.5.$$

The appropriate coding in these examples is simply the substitution of 0 for H and 1 for T. The codewords (keeping the same order) for the above mentioned values of ξ are 00, 11, 010, 011, 100, 101 and the expected length of a codeword is $\frac{1}{2} \cdot 2 + \frac{1}{2} \cdot 3 = 2.5$.

It follows from the law of large numbers that when ξ is observed sufficiently often and the result is coded as above, the average number of symbols needed (each of which is a 0 or 1) in any one observation will be less (with probability arbitrarily close to one) than $2.5 + \varepsilon$ for any arbitrarily small positive ε.

In these two examples, it was easy to find the optimal method of coding because $p_1, p_2, ..., p_N$ were powers with integer exponents of $\frac{1}{2}$ permitting us, at every step, to divide the still available possibilities into two classes of exactly equal probabilities. In the general case, this is not possible and it is more difficult to show the validity of the Shannon formula, although the essence of the proof remains the same, namely the law of large numbers.

Our lecturer gave us only the skeleton of the proof in the general case. He left it to us to fill out the details. The steps are as follows: if the random variable ξ assumes the value $x_1, x_2, ..., x_N$ with respective probabilities $p_1, p_2, ..., p_N$ and we observe ξ sufficiently often, — let's say n times, in such a way that the observations are independent, then according to the well-known rule of probability theory that the probability of the joint occurrence of independent events is equal to the product of the probabilities of the individual events, the probability of an outcome having occurrences n_1, of x_1, n_2 of $x_2, ..., n_N$ of x_N is $p_1^{n_1} \cdot p_2^{n_2} ... p_N^{n_N}$. According to the law of large numbers, if δ and ε are arbitrary small positive numbers and n is large enough, then $\frac{n_1}{n}$ will differ, (with a probability $(1-\delta)$) from p_1 by less than ε, ..., $\frac{n_2}{n}$ from p_2 by less than ε, ... and $\frac{n_N}{n}$ from p_N by less than ε, and the probability given above will be approximately $q=(p_1^{n_1}, p_2^{n_2}, ..., p_N^{n_N})$. Since the sum of the probabilities of all possible outcomes is 1, each value of ξ (not counting the infrequent case having probability less than δ) will be one of the $\frac{1}{q}$ sequences and one will need approximately $\log_2 \frac{1}{q}$ number of 1's and 0's to code these sequences. But

$$\log_2 \frac{1}{q} = n \log_2 (p_1^{-p_1} p_2^{-p_2} ... p_N^{-p_N}) =$$

$$= n \left(p_1 \log_2 \frac{1}{p_1} + ... + p_N \log_2 \frac{1}{p_N} \right) = n H(\xi)$$

and so we proved that to code one outcome of ξ with 0's and 1's, the length of the needed codeword on the average is about $H(\xi)$, with a probability arbitrarily close to 1. Next, we proved the validity of the Shannon formula in general. Obviously, the Shannon formula contains the Hartley formula as a

special case, since if
$$p_1 = p_2 = \ldots = p_N = \frac{1}{N}$$
then
$$p_1 \log_2 \frac{1}{p_1} + p_2 \log_2 \frac{1}{p_2} + \ldots + p_N \log_2 \frac{1}{p_N} = N\left[\frac{1}{N} \log_2 \frac{1}{\left(\frac{1}{N}\right)}\right] = \log_2 N.$$

We also proved that for a fixed N, $p_1 \log_2 \frac{1}{p_1} + \ldots + p_N \log_2 \frac{1}{p_N}$ will be, for every distribution $(p_1, \ldots, p_N)$ other than a uniform one, less than $\log_2 N$. This means that among the random variables that can have only N different values, the particular variable for which a given value will contain the most information is the one which assumes all of its values with equal probability.

The proof of this is as follows: as it is known that the curve $y = \log_2 \frac{1}{x}$ is convex from below, if we choose N arbitrary points on this curve and put positive masses on these points, the center of mass of these masses will always be above the curve, which means that

$$p_1 \log_2 \frac{1}{p_1} + p_2 \log_2 \frac{1}{p_2} + \ldots + p_N \log_2 \frac{1}{p_N} \leq$$
$$\leq \log_2 \left(p_1 \frac{1}{p_1} + \ldots + p_N \frac{1}{p_N}\right) = \log_2 N.$$

At the end of the lecture, the professor told us that both Shannon and Wiener arrived at the formula

$$H(\xi) = p_1 \log_2 \frac{1}{p_1} + p_2 \log_2 \frac{1}{p_2} + \ldots + p_N \log_2 \frac{1}{p_N}$$

in 1948, independently of each other. Actually, this formula had already appeared in the work of Boltzmann, which is why it is also called the Boltzmann–Shannon formula. Boltzmann arrived at this formula in connection with a completely different problem. Almost half a century before Shannon, he gave essentially the same formula to describe entropy in his investigations of statistical mechanics. He showed that if, in a gas containing a large number of molecules, the probabilities of the possible states of the individual molecules are $p_1, p_2, \ldots, p_N$, then the entropy of the system is $H = c\left(p_1 \log \frac{1}{p_1} + p_2 \log \frac{1}{p_2} + \ldots + p_N \log \frac{1}{p_N}\right)$, where c is a constant. (In statistical mechanics the natural logarithm is used and not the base 2, but this doesn't matter since the two are equal to a constant factor.) The entropy of a physical system is the measure of its disorder. One can also think of it as a quantity characterizing the uncertainty of the state of the molecules in that system.

Given this interpretation, we can easily see why Boltzmann arrived at the same formula for entropy as Shannon and Wiener did for information.

After some thought, one can see that uncertainty is nothing but a lack of information, i.e., negative information. In other words, information is the decrease of uncertainty. Before we observe a value of ξ, we are uncertain about which of its possible values ξ is going to have. After the observation of ξ, this uncertainty is removed. Now we know that this observation contains $H(\xi)$ bits of information. This information removes the uncertainty about the value of ξ which existed before the observation, therefore, it is reasonable to choose the number $H(\xi)$ for the measure of uncertainty. $H(\xi)$ can then be looked upon as the measure of the uncertainty which exists with respect to the value of ξ before it is determined. The measure of uncertainty is called entropy. The Shannon formula can accordingly also be interpreted as follows: if the random variable ξ assumes values $x_1, x_2, ..., x_N$ with respective probabilities $p_1, p_2, ..., p_N$, then the entropy of ξ denoted by $H(\xi)$, can be calculated by the formula $H(\xi) = \sum_{k=1}^{N} p_k \log \frac{1}{p_k}$.

At this point, we asked the professor why entropy and information are denoted by the letter H. He answered that it was Boltzmann himself who had introduced this notation and after that, it became customary. He also pointed out that the entropy $H(\xi)$ of a random variable ξ (or the amount of information contained in the observation of ξ which amounts to the same thing) is independent of the values of ξ, i.e., $x_1, x_2, ..., x_N$ (about which it is enough to know that they are different numbers) but depends on the probabilities — $p_1, p_2, ..., p_N$ — with which ξ assumes these values. If $f(x)$ is a function such that its values are different at the arguments $x_1, x_2, ..., x_N$, then the entropy of the random variable $f(\xi)$ is the same as the entropy of ξ, $H[f(\xi)] = H(\xi)$. We also proved by an application of the convexity of the function $x \log_2 x$, that if the values of $f(x)$ at $x_1, x_2, ..., x_N$ are not all different, then $H[f(\xi)] < < H(\xi)$ which means that the uncertainty regarding the value of $f(\xi)$ is less than that of ξ.

We then extended the law of additivity of information to the general case. This law can be stated in relation to the general case as follows: if ξ and η are independent random variables, then the information content of the simultaneous observation of ξ and η, denoted by $H((\xi, \eta))$ is equal to the sum of the information contained in the individual observations of ξ and η, i.e.,

$H((\xi,\eta)) = H(\xi) + H(\eta)$. This relation follows from the basic laws of logarithms. To give this in more detail, if ξ can have the value x_k ($k=1, 2, ..., N$) with probability p_k and η can have the value y_j ($j=1, 2, ..., M$) with probability q_j, then the pair (ξ, η) of variables will assume the value (x_k, y_j) with probability $p_k q_j$. Using this fact and also $\sum p_k = \sum q_j = 1$, we have:

$$H((\xi,\eta)) = \sum_{k=1}^{N} \sum_{j=1}^{M} p_k q_j \log_2 \frac{1}{p_k q_j} = \sum_{k=1}^{N} \sum_{j=1}^{M} p_k q_j \left(\log_2 \frac{1}{p_k} + \log_2 \frac{1}{q_j} \right) =$$

$$= \left(\sum_{j=1}^{M} q_j \right) \left(\sum_{k=1}^{N} p_k \log_2 \frac{1}{p_k} \right) + \left(\sum_{k=1}^{N} p_k \right) \left(\sum_{j=1}^{M} q_j \log_2 \frac{1}{q_j} \right) =$$

$$= H(\xi) + H(\eta).$$

If I wanted to be malicious I'd say that what we learned in the second lecture was that nothing of what we had learned in the first was true. Of course, this is not correct; we were only giving greater precision to what we had formulated inexactly on the first occasion. It seems to me that the professor's method is to explain the difficult notions in several stages. In his first approximation, he neglects certain things of secondary importance in order to put more emphasis on the essence of the matter; later, he patches up what was not very precise. This method of introducing a new concept is similar to the way a sculptor carves a statue out of a piece of marble or wood. First, he cuts out just the outline and later he comes back to the finer details. Although it is unusual, there are certain advantages to this method, the main one being that it forces one to do independent critical thinking. The usual system of taking notes doesn't work with this kind of teaching, since the notes taken at earlier lectures have to be corrected after the later ones. If I used the usual note-taking procedure (writing down only the definitions, theorems and their proofs) it would cause me difficulties; so I have another reason to be glad that I decided on this diary-form of lecture-notes.

Let's have a closer look at what needs to be corrected in what was heard at the first lecture. First of all, the statement that every "yes" or "no" answer (or signal which can have only two values, let's say 0 and 1) always contains 1 bit of information. According to what we heard at the second lecture, this should be modified to state that such a signal contains 1 bit of information only if the two possible values are equally probable — i.e., both have probability $\frac{1}{2}$ — otherwise, its information-content is less than 1 bit. To put this more precisely: if, for example, we ask a question in the Bar-kochba game to which the "yes" answer has probability p and therefore the "no" has probability $(1-p)$, then according to the Shannon formula the answer contains

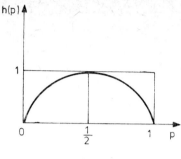

Fig. 1

$h(p) = p \log_2 \dfrac{1}{p} + (1-p) \log_2 \dfrac{1}{(1-p)}$ bit information. I have drawn the $h(p)$ function (Fig. 1). The curve reaches the value 1 at $p = \dfrac{1}{2}$ only; elsewhere, it is less than 1 and is symmetrical to the vertical line drawn through $p = \dfrac{1}{2}$. This latter fact can also be seen from the formula, since $h(p) = h(1-p)$, so $h\left(\dfrac{1}{2}+x\right) = h\left(\dfrac{1}{2}-x\right)$ if $0 \leq x \leq \dfrac{1}{2}$. This should hold because there cannot be any difference whether it is the yes or the no answer to a certain question that has probability p. If, for example, somebody throws a die and I ask him "Did you get a 6?", then, since the probability of a "yes" answer is $\dfrac{1}{6}$, the answer to my question contains $h\left(\dfrac{1}{6}\right) = \dfrac{1}{6}\log_2 6 + \dfrac{5}{6}\log_2 \dfrac{6}{5} = 0.65$ bit of information, which is less than 1 bit.

Or, to go back to the lecturer's joke, if I ask a girl whether she wants to marry me, the information content of the answer depends on the probability with which she will say yes. If this probability is very small or else very close to one, then the answer will not contain too much information — I will hear what I know almost with certainty. When I got to this point, it suddenly became clear what the lecturer meant by remarking that if ξ assumes its possible values with different probabilities, then in trying to guess its actual value, one should ask each question so that a "yes" answer has a probability as close to $\dfrac{1}{2}$ as possible. In this way, one can get the greatest amount of information out of each answer. He could have told us this, but he wanted us to realize it for ourselves. I can even calculate how much information I get when I ask questions clumsily. If I have to guess one of the first eight natural numbers, all of them being equiprobable with probability $\frac{1}{8}$, and if I ask first whether the number is one of 1, 2, 3, 4, then I will get 1 bit of information, since the probability of a "yes" answer to this is $\frac{1}{2}$. If, as a second question I ask whether the number is one of 1, 5, 6, 7, then having a "yes" answer to the first question will make the probability of a "yes" to the second one $\frac{1}{4}$ (conditional). If the answer to the first question was "no", then I will get a yes answer, with (conditional) probability $\frac{3}{4}$. In both cases, the second answer contains only $\left(\frac{1}{4}\right) = 0.83653$ of a bit of new information as 0.16347 of a bit was already known. There is only 0.83653 of a bit of information that is new out of its 1 bit of information content.

Having learnt this, I returned to the false coin problem. I realized that it is possible to determine the false coin (which is lighter than the other 26) with 3 measurements only because the measurements can be chosen in such a way that their outcomes have equal probabilities. If the measurements have 3 possible outcomes but not with equal probabilities, then one measurement supplies less than $\log_2 3$ of a bit of information. Let's say there are 25 coins, one of them being false. With 8–8 coins put in the two pans of the scales, the probabilities of the three outcomes are $\frac{8}{25}, \frac{8}{25}, \frac{9}{25}$ and by the Shannon formula, the experiment will contain $2 \cdot \frac{8}{25} \log_2 \frac{25}{8} + \frac{9}{25} \log_2 \frac{25}{9} = 1.58269$ bits of information instead of $\log_2 3 = 1.58496$ bits.

I thought a lot about what it means that information can be interpreted as a decrease of uncertainty. It is true that when a guessing game starts, I am in complete uncertainty as to what I'm going to guess. As the game progresses, this uncertainty decreases with the answers to my questions and it disappears when I have guessed correctly. If this uncertainty was B bits at the beginning, then after receiving x bits of information, by definition, there are still $B-x$ bits of uncertainty.

Therefore at any given time during the game, the sum of the existing uncertainty and the information so far collected are constant, since $x+(B-x)=B$. This relation looked very familiar; I knew I had already met with it. After some thought, I realized how similar it is to the constant sum of the kinetic and potential energies during free fall.

When a brick lies on a roof it has only potential energy, and no kinetic energy. When it starts to fall, its kinetic energy increases and the potential energy decreases, so that the sum of the two during the fall is constant. This analogy reveals a very exciting problem: — *it seems there is some similarity or analogy between the concepts of information and energy.* Moreover, it looks as though there exists a law which can be called the law of conservation of information. This connection can be stated in another way, too: in Barkochba, the sum of the information already received and that which is still missing is constant. I decided to ask the lecturer at the next class if hypothesis of a parallel between the concepts of information and energy is correct.

I see yet another similarity: between energy-conversion (for example, electrical energy into mechanical energy, etc.) and information coding. The TV transmitter codes the information in a picture into information in electromagnetic waves; the TV receiver converts this information into a picture again. This process reminds me very much of the conversion of mechanical energy into electrical (in a generator) and after the transmission, another conversion into mechanical energy in an electric motor. I have the feeling that there is a basic relation between energy and information: I am very eager to follow this up.

Third lecture

The first idea we looked at today was that of conditional entropy (or information). If B is an arbitrary event of positive probability, ξ is a random variable having $x_1, ..., x_N$ possible values (all being distinct), and A_k is the event that ξ_k assumes the value x_k ($k=1, 2, ..., N$), then the conditional entropy of ξ, given condition B, is defined as the entropy of ξ's conditional distribution given B, i.e.,

(1) $$H_B(\xi) = \sum P(A_k|B) \log_2 \frac{1}{p(A_k|B)},$$

where $P(A_k|B)$ is the conditional probability of event A_k given B, $P(A_k|B) = \frac{P(A_k B)}{P(B)}$, where $A_k B$ denotes the event that A_k and B happen together.

If we now take another random variable η, which can assume the values $y_1, ..., y_M$ and B_j denotes the event $\eta = y_j$ ($j=1, 2, ..., M$), then the conditional entropy of ξ given a certain η, denoted by $H_\eta(\xi)$, is defined as the expected value of $H_{B_j}(\xi)$:

(2) $$H_\eta(\xi) = \sum P(B_j) H_{B_j}(\xi),$$
$$= \sum\sum P(A_k B_j) \log_2 \frac{P(B_j)}{P(A_k B_j)}.$$

Let us consider how much the entropy of ξ, decreases uncertainty about the value of ξ, by observing η. This quantity, which we will denote by $I(\xi, \eta)$, can be considered as the amount of information concerning ξ by observing η. Therefore, by definition:

(3) $$I(\xi, \eta) = H(\xi) - H_\eta(\xi) = \sum P(A_k) \log_2 \frac{1}{P(A_k)} -$$
$$- \sum \sum P(A_k B_j) \log_2 \frac{P(B_j)}{P(A_k B_j)}.$$

Since $\sum_j P(A_k B_j) = P(A_k)$ (which is true because the events $A_k B_1, \ldots, A_k B_n$ are mutually exclusive, and if A_k occurs, then only one of events $A_k B_j$ can occur), it follows that:

(4) $$I(\xi, \eta) = \sum \sum P(A_k B_j) \log_2 \frac{P(A_k B_j)}{P(A_k) P(B_j)}.$$

We reached the following conclusions about $I(\xi, \eta)$.

a) $I(\xi, \eta)$ is always non-negative and is zero only if ξ and η are independent. (This follows from the convexity of the function $\log_2 \frac{1}{x}$). If ξ and η are independent, then the observation of η will not supply any information concerning ξ. If they are not independent, then the observation of η will contain some information about ξ, too. The professor joked that from this we can conclude that whatever we learn at the University, we can only end up smarter and not stupider since in the worst case, it will only be a zero amount of information that we get out of our studies.

b) $I(\xi, \eta) \leq H(\xi)$ where the equality sign holds if and only if $\xi = f(\eta)$, i.e. if the value of η determines the value of ξ uniquely, when we can get the exact value, i.e., full information of ξ by observing η (which will dissolve the $H(\xi)$ uncertainty about ξ completely). In particular, this is the case when $\eta = \xi$, i.e., $I(\xi, \xi) = H(\xi)$.

c) $I(\xi, \eta) = I(\eta, \xi)$, meaning that the observation of η gives as much information about ξ as the observation of ξ about η. That is why $I(\xi, \eta)$ is usually called the *relative information of ξ and η*.

The lecturer made a very interesting theoretical comment about this last characteristic. He said that the deep cause of the equality of $I(\xi, \eta) = I(\eta, \xi)$ is as follows: if we investigate two entities that are random and to a certain extent dependent on each other, then we cannot by using information theory deduce which of the two is the cause and which is the effect in their relationship. The only thing that can be established is how close their dependence is. Let's denote the water-level of the river Danube in Budapest on any day of the year by η and the amount of precipitation during the previous week in Bavaria by ξ. Obviously, although ξ and η are random, there is some connection between the two, namely, that if there is a lot of rain in Bavaria, the water-level of the Danube will rise in Budapest. This dependence is a causal one, because it is the rain in Bavaria that causes the elevation of the water-level of the Danube in Budapest and not vice-versa. Although this is not a deterministic relationship because there are several other parameters on which the water-level of the Danube depends (such as the amount of rain in Austria, Slovakia, the Transdanubia, etc.), still, there is a definite dependence. Therefore the observation of ξ gives some information about η and vice versa, but since the information content of the two are equal, we cannot conclude anything about the causal relationship between ξ and η.

The lecturer suggested that those who are not afraid of complex calculations try the following problem: assume that we take an n-element sample from a conveyor-belt carrying N parts at a time. Let ξ denote the percentage of faulty items among all of the parts and η denote the percentage of faulty parts in the sample. Let us see how much information about ξ is obtained by observing η. By calculating this information, we will be able to establish some guidelines on how large a sample needs to be to provide sufficient information about the percentage of faulty elements on the whole.

In class we calculated the answer to the following problem: there are two urns, one of which has a red and b white balls, and the other b red and a white. The two urns are exactly the same: by looking at them one cannot tell them apart. Let's choose one of the urns and take a ball from it. Let $\xi=1$ or $\xi=2$ if we choose the first or the second urn, respectively. Let $\eta=1$ if the ball is red and $\eta=2$ if it is white. Calculate $I(\xi, \eta)$, i.e., determine the amount of information supplied by the color of the chosen ball about which urn it came from. Let A_1 denote the event when $\xi=1$ and A_2 that when $\xi=2$; similarly, let B_1 denote the event when $\eta=1$ and B_2 that when $\eta=2$. Then:

$$P(A_1) = P(A_2) = P(B_1) = P(B_2) = \frac{1}{2},$$

$$P(A_1 B_1) = P(A_2 B_2) = \frac{a}{2(a+b)},$$

$$P(A_1 B_2) = P(A_2 B_1) = \frac{b}{2(a+b)}$$

and so,

$$I(\xi, \eta) = \frac{a}{a+b}\log_2\frac{2a}{a+b} + \frac{b}{a+b}\log_2\frac{2b}{a+b} = 1 - h\left(\frac{a}{a+b}\right).$$

Therefore if, for example, $a=1$ and $b=3$, then $I(\xi, \eta)=1-h\left(\frac{1}{4}\right)=0.16347$, while if $a=1$ and $b=7$, then $I(\xi, \eta)=1-h\left(\frac{1}{8}\right)=0.4564$. (When $a=b$, then $I(\xi, \eta)=0$, and ξ and η are independent.) The result, in other words, is that the desired bit of information concerning ξ is approached (by the observation of η) better when $\frac{a}{a+b}$ is farther from $\frac{1}{2}$.

Then the lecturer discussed another way to get at the concept of entropy or relative information. Let us associate with every event a number $V(A)$ which represents the *unexpectedness of event A*. We assume $V(A)$ to have the following characteristics.

1. $V(A)$ depends only on the probability of event A; it is a function of it, i.e.,

(5) $$V(A) = f[P(A)],$$

when $f(x)$ is a monotone decreasing function (meaning that the more improbable an event is, the greater the unexpectedness of its occurrence).

2. If events A and B are independent, then the unexpectedness of their simultaneous occurrence is equal to the sum of their individual unexpectedness. Denoting by AB the event that A and B occur simultaneously, we have:

(6) $$V(AB) = V(A)+V(B).$$

3. We choose, as the unit of unexpectedness, the unexpectedness of an event having probability $\frac{1}{2}$ (for example the tossing of a coin), i.e.,

(7) $$f\left(\frac{1}{2}\right) = 1.$$

These three assumptions hold if

(8) $$V(A) = \log_2 \frac{1}{P(A)}.$$

It can readily be seen (using the property of the logarithm-function — as a monotone function — that the logarithm of a product is equal to the sum of the logarithms of the factors) that no other function can satisfy the above three assumptions. The unexpectedness $V(A)$ of a random event A is defined by (8).

We can now define the *entropy* of a random variable (having only countable many values) as the expected value of the unexpectedness of the value assumed by the variable. If the random variable ξ has the possible values $x_1, x_2, ..., x_N$

with probabilities $p_1, p_2, ..., p_N$, respectively, A_k denotes the event that ξ assumes the value x_k ($k=1, 2, ..., N$). The unexpectedness of A_k is $\log_2 \frac{1}{p_k}$ and the probability of A_k is p_k, therefore the entropy of ξ is

(9) $$H(\xi) = p_1 \log_2 \frac{1}{p_1} + p_2 \log_2 \frac{1}{p_2} + ... + p_N \log_2 \frac{1}{p_N}.$$

We have arrived at the Shannon formula again, but in a different way. This way also brings us to the concept of *relative information*.

Let A and B be arbitrary event relating to the same experiment. If we observe the outcome of event B, this will change the unexpectedness of event A. The unexpectedness of A was originally $\log_2 \frac{1}{P(A)}$. After the observation of B, the probability of A has changed to the conditional probability $P(A|B) = \frac{P(AB)}{P(B)}$ and its (conditional) unexpectedness will be $\log_2 \frac{1}{P(A|B)}$. Let $V(A, B)$ denote the change in the unexpectedness of A resulting from the observation of B, then

(10) $$V(A, B) = \log_2 \frac{1}{P(A)} - \log_2 \frac{1}{P(A|B)} = \log_2 \frac{P(A|B)}{P(A)} =$$

$$= \log_2 \frac{P(AB)}{P(A)P(B)}.$$

$V(A, B)$ is positive if $P(AB) > P(A)P(B)$; it is negative if $P(AB) < P(A)P(B)$, and it is 0 if $P(AB) = P(A)P(B)$, i.e., when A and B are independent. If A and B are independent, then the observation of B will not change the unexpectedness of A, while in the case where is dependence, the observation of B will decrease or increase the unexpectedness of A depending on which of $P(AB)$ and $P(A)P(B)$ is greater.

Now, let ξ and η be two random variables; let ξ have possible values $x_1, x_2, ..., x_N$ and η have possible values $y_1, y_2, ..., y_M$. Let A_k be the event where $\xi = x_k$ ($k=1, 2, ..., N$) and B_j be the event where $\eta = y_j$ ($j=1, 2, ..., M$). We want to see by how much, on the average, the value of the unexpectedness of ξ will change with the observation of η. In other words, we want to calculate the expected value of $V(A_k, B_j)$. We obtain:

(11) $$\sum_{k=1}^{N} \sum_{j=1}^{M} P(A_k B_j) V(A_k B_j) = \sum_{k=1}^{N} \sum_{j=1}^{M} P(A_k B_j) \log_2 \frac{P(A_k B_j)}{P(A_k) P(B_j)} = I(\xi, \eta).$$

The result, then, is the following: the relative information $I(\xi, \eta)$ can be defined as the expected change in the unexpectedness of ξ brought about by the observation of η. The two definitions of $I(\xi, \eta)$ are equivalent, and this can be expressed by stating that the decrease in the uncertainty of a given random variable ξ by the observation of a random variable η is equal to the expected change in the unexpectedness of the value of ξ resulting from the observation of η.

After the lecture, I asked the lecturer if my conclusion about the analogy between the notions of information and energy was correct. He was very pleased with my question and indicated that the parallel was correct. He also said that he had wanted to talk about this analogy himself, but now that I discovered it, it would be better for me to prepare a talk on it. He promised to give me the necessary bibliography on this question at the next lecture. So my question back-fired but I don't really mind. The problem is that much on my mind, although it is not a mathematical question but a philosophical one. I have always been interested in philosophical problems relating to mathematics.

Today's lecture made me think of the difference and the similarity between the concepts of uncertainty and unexpectedness. In my opinion, the lecturer passed over the underlying logical difficulties much too quickly, possibly because he had gotten so used to these ideas, he could not see what difficulties they can cause a beginner. It was difficult enough but I think I have it clear now. To begin with, one should realize that these two words refer to two different things: a random event has unexpectedness, while a random variable has uncertainty. The difference can best be shown by investigating an arbitrary random event A together with the random variable α, the values of which depend only on whether A will happen or not. Let's say $\alpha=1$ if A occurs and $\alpha=0$ if it doesn't. (In other words, the random variable α is the *indicator* of the event A.) Let the probability of A be $P(A)=p$. The unexpectedness of the event A is $V(A)=v(p)=\log_2 \frac{1}{p}$, while the uncertainty (entropy) of the indicator α of the recent A is $h(p)=p\log_2\frac{1}{p}+(1-p)\log_2\frac{1}{1-p}$ (Fig. 2). From Fig. 2 it can be seen that the $h(p)$ and $v(p)$ functions are equal for only two p values: if $p=\frac{1}{2}$, then $h(p)=v(p)=1$ and if $p=1$, then $h(p)=v(p)=0$. If p approaches zero $h(p)$ goes to zero while $v(p)$ increases without limit. More light can be shed on this by arriving at the following relationship: $h(p)=pv(p)+(1-p)v(1-p)$, which means that the uncertainty concerning α is a weighted (by the probabilities of the corresponding events) average of the unexpectedness of the events A and $\bar{A}$ (i.e., events $\alpha=1$ and

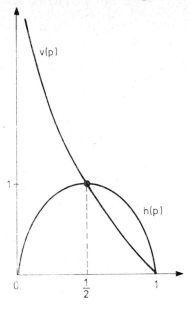

Fig. 2

$\alpha=0$). The uncertainty of α is greatest when $p=\frac{1}{2}$, while the unexpectedness of the event A is greater and greater as p is smaller and smaller.

Let us compare the decrease in uncertainty with a change in unexpectedness. The situation is like this: With the observation of a random variable, the uncertainty (entropy) relating to another random variable will always decrease, or stay as it is — the latter result occurring in the case of two independent random variables. On the other hand, with the observation of an event, the unexpectedness of another event can decrease, increase or stay the same — the last possibility again occurring in the case of independent events. A decrease in uncertainty can always be interpreted as information. A change of unexpectedness is not information; only its expected value can be taken for information (because the expected value of the unexpectedness is equal to the decrease of uncertainty).

I also thought about what the "dimension" of unexpectedness might be, in other words, if it can be expressed in bits. The way I see it, this is possible, from a purely mathematical point of view, but it would be improper because it would mean treating the change in unexpectedness as an amount of information, which it obviously is not! (One can see that the identification of a change in unexpectedness with information is incorrect by realizing that it would mean that, in certain cases, an observation would provide negative information — and this is obviously absurd.) It would be possible to measure unexpectedness in some new kind of unit and give it a name too, but there is no need for all that. Unexpectedness doesn't need "dimension", it is enough to make it a dimensionless number.

And I thought about the lecturer's joke too, that from $I(\xi, \eta) \geq 0$, it follows that no matter what we learn at the University it cannot harm us since in the worst case, it will simply be of no use. Of course, one can say that one will get smarter by studying and thinking, but simple memorizing makes the mind dull, so I can't say that preparing for exams doesn't destroy one's intellectual capacities. There has been a lot of talk lately about the need to decrease the number of exams. I really don't think that the problem is their number, but much more the way they are conducted. We need the kind of exam where the requirement is not to regurgitate what we have desperately tried to jam into our brains in the previous couple of days, which will be mostly forgotten during the preparation for the next exam. Rather, we need the kind of exam which will allow us to show our understanding of the material and to demonstrate our thinking abilities. Of course, I have no detailed idea of the how and what of such an exam — it is possible that it's just an impossible dream of mine. Or is it possible that information theory can help to answer this question? Because it is information that this is all about; the examiner must get information about how much information the student has accumulated on the given subject.

Once, when we were discussing the exam timetable with the professor, he remarked that he considered our way of thinking about education at the University to be too "exam-centered" and that he disagreed with such thinking. He told us a story about Niels Henrik Abel, the great Norwegian mathematician, who as a young student went to Berlin to pursue his studies. He knew hardly any German at the time, so when he reported to Professor Crelle, he said only a few words and gave Crelle the letters of recommendation from his teachers in Oslo. Crelle assumed that Abel wanted some information about the exams, so he explained at length when and in what subjects he would have to take them. After a while, Abel, who already had with him in his briefcase at that time more than one paper containing quite important new findings, interrupted Crelle and said to him in broken German: "Nee, Herr Professor, nix Examen, nur Mathematik!" Our professor said that he missed that kind of spirit in today's students. One had to admit he had a point. Most students are exam-centric. But I doubt that students alone are to blame. The root of the problem lies in the present educational system — the exam-centricity originates with the university, we are just influenced by it. Whenever an evaluation is made of the work of individual students, or of various groups of students, for instance, for the purpose of selecting those who will receive scholarships or grants, the only criterion used in the evaluation seems to be the exam grades of those involved.

This is what needs to be changed. As far as students in Mathematics are concerned, one solution could be to have them give seminars, do assignments, write papers and have them evaluated on such work. It would seem possible to put such a system into effect quite easily for mathematicians; of course, I don't know that much about the situation in other Faculties. But why should the procedure be the same everywhere, in all Faculties of the University? This mechanical uniformity or the tendency towards it is one of the main problems. Since the problems manifest themselves differently from Faculty to Faculty the solutions should be different, too.

Can the difficulty of an exam be measured by how many bits of information a student would need to pass it? This may not be so absurd in the encyclopedic subjects but in mathematics it doesn't make any sense since things follow from each other and, in principle, whoever knows the bases knows everything. All of the results of a mathematical theorem are in the axioms of mathematics in embryonic form, aren't they? *I will have to think this over some more.*

Returning to the inequality $I(\xi, \eta) \geqq 0$ and the lecturer's remark about the effects of our studies, made me aware that he knows how much we have argued among ourselves over the past few weeks as to whether we would really learn what we need most in our work after our graduation. Many of us had definite doubts about this. But they couldn't agree on which subjects were unnecessary because each of them thought that those in which he had an interest and which he studied with pleasure were the important ones. Those which were considered unnecessary turned out to be those that a given individual wasn't too interested in or those in which he had difficulties — therefore almost all subjects had their defenders and attackers. Theoretically I didn't agree with all this criticism, since I considered even its starting point faulty. I am quite aware of the kind of jobs taken by graduates in the recent years, and I see clearly that there will be as many kinds of jobs as there are graduates. So it is impossible to figure out what specialized knowledge one will need. We have to face the fact that we won't learn everything we need later in our work at the University: we will have to supplement our knowledge with independent study. And therefore we should consider the subjects in our curriculum from the viewpoint of how much help they will provide for those further independent studies. In that respect, it is not only the subject itself that will be useful to us but also the rational method that will be learned and the ability to think that will be acquired. Now then, it is impossible for us to know beforehand how much help the abilities we develop will be in the study of a given subject of which we know nothing more at the moment than its name. Therefore, our arguments about the curriculum are pointless. My opinion was at first very unpopular among my classmates: they argued that it was wrong for me to disclaim the right of university students to seek to influence the curriculum since the trend in universities these days is very much toward a policy of restriction in any case.

My answer to that was that I don't want to disclaim our rights. I merely suggest that we voice our opinions only on those issues about which we are qualified to give opinions. For example, it is a fact that we can judge best whether it is worthwhile to go to a certain class, or if a tutoring session is useful or not, because we can see if these help us in our studies or not. And we can clearly see if an examiner is unjust in his marking. But we lack the necessary overview to talk about the curriculum.

I guess when we got into an argument about the teaching of physics, my classmates finally understood my point of view and that I don't want to give up our right to critize the quality of teaching if necessary. What happened was that I was the one who criticized the present situation most strongly, probably because I am very much interested in physics. As well as indicating what mathematics majors dislike in the present method of teaching physics, I also described how I thought it should be taught. While realizing — rightly — that the system is basically useless in its present form, most of my colleagues had never even given a thought to the possibility that there might be other ways of teaching physics. Their conclusion was that physics should not be taught to mathematics students at all or at most only in limited amounts. They even philosophized that while physics had been historically the main area in which mathematical methods had been applied, the center of application had now shifted to economics. Although there is some truth to this argument, the reasoning is not correct. It is still true that physics is the main field of application of mathematics. This is true not only of traditional areas of mathematics but even of many new ones (such as functional analysis, group theory, complex functions of one or several variables, distribution theory, etc.). As a matter of fact, physics still constitutes the major inspiration for new developments in mathematics. So, in my view, the number of lectures on physics for mathematics students is by no means too great, but a different approach is needed. At present, our physics lectures are the same as those for the physics students and that is not appropriate. We need physics courses where the main stress is on the use of mathematical methods. One way to accomplish this, in my opinion, would be to have the lectures given not by physicists but by mathematicians knowledgeable concerning the *applications of mathematics in physics.*

THIRD LECTURE

This would answer our main complaint in this area namely that the lecturers in physics usually treat the question of mathematical precision quite offhandedly. They claim that the scientific insight will prevent us from reaching false conclusions, merely because of a lack of precision in the mathematics used. Of course, this is not at all soothing to us, since we are interested in the application of mathematics to physics! Recently I got into a real argument with one of the instructors in physics when I mentioned to him — in connection with his lack of mathematical precision — what we had learned in mathematical logic, namely that from one false statement any other true or false statement can be derived. And since he had written down an obviously false equation, he didn't need to bother doing anything more, since anything could be deduced from it.

Returning to what we had discussed in the information theory class, I thought some more about the $I(\xi, \eta) = I(\eta, \xi)$ connection and the direction of cause–effect relations. My impression is that our lecturer simplified the question a bit. From $I(\xi, \eta) = I(\eta, \xi)$, it doesn't follow that $I(\xi, \eta)$ gives information only about the strength and not about the quality of the connection between ξ and η. Toss one coin twice and denote the result by ξ. The possible values of ξ are HH, HT, TH, TT, where H stands for heads and T for tails. Let $\eta = 0$ if the results of both tosses are the same (both are either heads or tails), and let $\eta = 1$ if the results are different (one is a head, the other is a tail). In this case, the value of ξ uniquely determines the value of η but the reverse is not true. Therefore, $I(\xi, \eta) = H(\eta) = 1$ while $I(\xi, \eta) \neq H(\xi) = 2$. Although it is still true that $I(\xi, \eta) = I(\eta, \xi)$, one can see even from the amounts of information alone that the value of ξ determines the value of η but that the value of η doesn't determine the value of ξ [since $I(\xi, \eta) = H(\eta)$ but $I(\eta, \xi) \neq H(\xi)$]. In this example, the direction of the connection between ξ and η is evident and therefore the connection can be looked upon as causal (i.e., ξ is the cause, η is the effect). Now if neither ξ nor η is a function of the other, then the situation is more complicated, but I surmise by this example that one can derive something this way even in the general case. Again, I guess I will have to ask the professor, although it is possible that he himself will return to this question to make his observations more precise, as he has already done.

Fourth lecture

Today we analyzed mutual information further. One can see easily that

(1) $$I(\xi, \eta) = H(\xi) + H(\eta) - H(\xi, \eta).$$

From (1) and from the fact that mutual information is a non-negative entity, it follows that if ξ and η are random variables, then

(2) $$H((\xi, \eta)) \leq H(\xi) + H(\eta),$$

with equality if and only if ξ and η are independent. (1) can be written in the form:

(1') $$H((\xi, \eta)) = H(\xi) + H(\eta) - I(\xi, \eta) = H(\eta) + H_\eta(\xi).$$

(1') may be looked upon as the generalization of the law of additivity of information so that, if ξ and η are two arbitrary random variables, then, observing the value of ξ and η, we can get the information contained in these two observations if we add the information contained in the observation of η to the conditional information contained in an observation given a particular value of ξ. In other words, $H((\xi, \eta))$ can be obtained by subtracting from the sum of $H(\xi)$ and $H(\eta)$ the amount of information which is in it twice, namely $I(\xi, \eta)$. To make this is clear with an example: let α, β, γ be independent signs, each one of which can have the values 0 or 1 with probability $\frac{1}{2}$. Denote as ξ the (α, β) pair and as η the (β, γ) pair. Obviously, α, β and γ contain 1 bit of information each, and ξ and η contain two bits each, while the observation of the pair (ξ, η) is equivalent to the observation of α, β, γ (β twice but that is unimportant for the present) and so $H((\xi, \eta)) = 0$. Finally, $I(\xi, \eta) = 1$ since if we observe the outcome of ξ, then we know what values α and β have assumed, of which α gives no information about the independent η, while β supplies 1 bit (knowing β from the $\eta = (\beta, \gamma)$ pair, it is only γ which remains unknown). Since $3 = 2 + 2 - 1 - 2 + 1$, therefore, in this case, (1') holds.

We then began to examine another area of information theory, the information theoretical distance of two distributions.

Let

$$P = \{p_1, p_2, ..., p_N\} \quad \text{and} \quad Q = \{q_1, q_2, ..., q_N\}$$

be probability distributions consisting of the same number of positive elements (i.e. $\sum_{k=1}^{N} p_k = \sum_{k=1}^{N} q_k = 1$), then the information theoretical distance of distribution P from distribution Q, denoted by $D(P, Q)$ is defined by the following formula:

$$D(P,Q) = \sum_{k=1}^{N} p_k \log_2 \frac{p_k}{q_k}. \tag{3}$$

Because $\log_2 \frac{1}{x}$ is convex, it follows that $D(P,Q)$ is always non-negative and equal to zero only if the P and Q distributions are the same. If they are not, then among the terms on the right-hand side of equation (3) there are necessarily some which are positive and some which are negative, but as we have seen, the sum is always positive! $D(P,Q)$ can be written as

$$D(P,Q) = \sum p_k \left(\log_2 \frac{1}{q_k} - \log_2 \frac{1}{p_k} \right) \tag{3'}$$

and then we can interpret $D(P,Q)$ in the following way. Let the events $A_1, A_2, ..., A_N$ be the possible mutually exclusive outcomes of an experiment. Assume that distribution Q consists of the probabilities of these outcomes, i.e., $q_k = P(A_k)$ ($k=1, 2, ..., N$), while P consists of the probabilities of the outcomes of the same experiment under different circumstances. Then $\left(\log_2 \frac{1}{q_k} - \log_2 \frac{1}{p_k} \right)$ denotes the change in the unexpectedness of A_k which is due to the different experimental circumstances, $D(P,Q)$ is equal to the expected value of this change. When the expected value is calculated, we should use the probabilities corresponding to the changed conditions, i.e., we should weigh with the p_k's (and not with the old probabilities, the q_k's) the $\left(\log_2 \frac{1}{q_k} - \log_2 \frac{1}{p_k} \right)$ amount. With the help of the $D(P,Q)$ distance, the information $H(\xi)$ and $I(\xi, \eta)$ can be expressed as follows. Denote by P the probability distribution of the random variable ξ; then if ξ assumes the value x_k with probability p_k, let $P = \{p_1, p_2, ..., p_N\}$. Denote by Q the N-element uniform distribution, i.e., let $Q = \left\{ \frac{1}{N}, \frac{1}{N}, ..., \frac{1}{N} \right\}$. Then

$$H(\xi) = \log_2 N - D(P,Q). \tag{4}$$

(4) shows the already well-known fact that $H(\xi) \leq \log_2 N$.

Now let ξ and η be arbitrary random variables. Let ξ, η have as possible values $x_1, x_2, ..., x_N$ and $y_1, y_2, ..., y_M$, respectively. Denote by A_k the event $\xi = x_k$ ($k=1, 2, ..., N$) and by B_j the event $\eta = y_j$ ($j=1, 2, ..., M$). Denote by $R = \{P(A_k B_j)\}$ the point distribution of the random variables ξ and η, and by $P*Q$ the probability distribution $\{P(A_k)P(B_j)\}$. $P*Q$ is therefore the joint probability distribution of such two random variables ξ_1 and η_1, of which the distribution of ξ_1 is equal to the distribution of ξ, and that of η_1 to η's; moreover, ξ_1 and η_1 are independent (as opposed to ξ and η, which are generally not independent). Then

(5) $$I(\xi, \eta) = D(R, P*Q).$$

In other words, the mutual information $I(\xi, \eta)$ is equal to the information theoretical distance between the R distribution and the $P*Q$ distribution.

Next, we turned to codes with variable length codewords. Let ξ be a signal (i.e., a random variable) with possible values $x_1, \ldots, x_N$ having probabilities $p_1, p_2, \ldots, p_N$, respectively. First we looked at the simplest case, that of the so-called *binary codes* in which only the binary digits 0 and 1 are employed. Let us code the $x_1, x_2, \ldots, x_N$ values with different sequences of 0's and 1's (where we allow the length of the sequences to vary). Such a code is called a *prefix code,* if no code word is the prefix (contained fully as the beginning) of any other code word. The big advantage of such prefix codes is that the ends of the code words don't have to be marked. The code words can be written one after the other without separation since such a text can be decomposed into separate code words in only one way because of the prefix characteristic. In other words, the prefix code is uniquely decodable. There are other uniquely decodable codes which do not satisfy the prefix condition, for example, the code consisting of 0 and 0! code words. Next, we consider the following example. Let the values of ξ be the decimal digits and let the following code words composed of zeros and ones be assigned to them:

Decimal digit	Binary representation
0	0 0
1	0 1
2	1 0 0 0
3	1 0 0 1
4	1 0 1
5	1 1 0
6	1 1 1 0
7	1 1 1 1 0
8	1 1 1 1 1 0
9	1 1 1 1 1 1

It is easy to see that this is a prefix code. If we now write an arbitrary sequence of these codewords (such that the same codeword can appear in the sequence more than once) the resulting sequence of code symbols can be decomposed into code words in only one way and therefore can be decoded uniquely. For example, the sequence

1110101111110100000000110011111110101

can be decoded into code words only in the following way:

110|101|11110|1000|00|00|01|1001|111111|01|01

and this corresponds to the decimal digit sequence

6 4 7 2 0 0 1 3 9 1 1.

A prefix code is called primitive if it cannot be shortened, i.e., if the resulting sequences of code words are not a prefix code no matter what way we take symbols from the code words. It can be easily seen in the case of a primitive prefix code that, if we choose any sequence s of 0's and 1's which is none of the code words, then either there is no code word with the prefix s, or if there is, and we write a one or a zero at the end of it, (in both cases) the resulting sequence will be either a code word or the beginning of one.

A convenient graphical representation of a binary primitive prefix code can be obtained by representing each code word by a terminal node in a tree. Starting from the root of the tree, two branches will lead from every node which is not a terminal node. If we assign a 0 to every branch on the left and a 1 to those on the right, then the terminal nodes will uniquely correspond to a 0–1 sequence where the successive digits can be thought of as providing the climbing instructions (the choice between left- and right-hand branches at every node). In this way, the sequences corresponding to the terminal nodes will be the desired code words. Such a tree is called code tree. For example, the tree representing the previously described primitive code is:

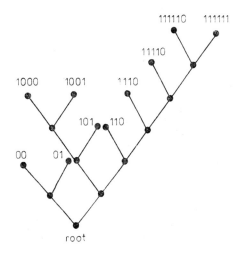

Let N_k denote the number of code words with length k in a primitive prefix code, and let the length of the longest code word be r. Then,

$$\text{(6)} \qquad \frac{N_1}{2^1} + \frac{N_2}{2^2} + \ldots + \frac{N_r}{2^r} = 1$$

holds. In our decimal to binary coding example: $N_1=0$, $N_2=2$, $N_3=2$, $N_4=3$, $N_5=1$, $N_6=2$, and so,

$$\frac{N_1}{2^1} + \frac{N_2}{2^2} + \frac{N_3}{2^3} + \frac{N_4}{2^4} + \frac{N_5}{2^5} + \frac{N_6}{2^6} = \frac{2}{2^6} + \frac{2}{2^2} + \frac{2}{2^3} + \frac{3}{2^4} + \frac{1}{2^5} + \frac{2}{2^6} =$$

$$= \frac{1}{2} + \frac{1}{4} + \frac{3}{16} + \frac{1}{32} + \frac{1}{32} = \frac{16+8+6+1+1}{32} = 1.$$

(6) can be proven in the general case as follows: take an arbitrary primitive prefix code and represent it with a code tree as described above. Let's imagine a monkey who climbs that tree, randomly choosing with probability $\frac{1}{2}$ between left and right branches at every branching point. If the monkey reaches a terminal node, it stays there. Obviously, the probability that the monkey will reach a terminal node to which a code word of length k corresponds (being at height k) is $\frac{1}{2^k}$. Since there are N_k such endpoints, the monkey will reach height k with probability $\frac{N_k}{2^k}$. But the monkey stops climbing only when it reaches a terminal node, therefore it must be true that $\sum \frac{N_k}{2^k} = 1$ (since the sum of the probabilities corresponding to the mutually exclusive outcomes of an experiment is always equal to 1).

(6) can be written in another form, too: if ξ assumes the values $x_1, x_2, \ldots, x_N$ and the appropriate primitive prefix code words composed of 0's and 1's have lengths of $l_1, l_2, \ldots, l_N$ respectively, then

$$\text{(7)} \qquad \frac{1}{2^{l_1}} + \frac{1}{2^{l_2}} + \ldots + \frac{1}{2^{l_N}} = 1,$$

because on the left side of (7) there are N_k of $\frac{1}{2^k}$, so that side is equal to $\sum \frac{N_k}{2^k}$ which, according to (6), is equal to 1. The numbers $\frac{1}{2^{l_1}}, \ldots, \frac{1}{2^{l_N}}$ therefore create a probability distribution which we will denote by Q. Let's calculate the information theoretical distance $D(P, Q)$ between $P = \{p_1, \ldots, p_N\}$ and Q.

Since this distance is non-negative, it is true that

$$D(P, Q) = \sum p_k \log_2 \frac{p_k}{\left(\frac{1}{2^{l_k}}\right)} \geqq 0,$$

(8) $$\sum p_k l_k \geqq H(\xi).$$

The meaning of $L = \sum p_k l_k$ is obviously the expected length of the code word of 0's and 1's assigned to ξ. Therefore, (8) means that if an arbitrary sign sequence is coded into sequences of 0's and 1's so that the resulting code is a primitive prefix code, the average code-word length cannot be smaller than the information content of ξ. The reason for this is clearly that a 0 or 1 sign can contain a maximum of 2 bits of information. Therefore, to code $H(\xi)$ bits with a sequence of zeros and ones the length of the sequence on the average should be at least $H(\xi)$, assuming that no information is lost. In the case of prefix codes, information cannot be lost since, as we have seen, these codes are uniquely decodable. In other words, inequality (8) expresses the principle of information conservation, meaning that $H(\xi)$ bits cannot be compressed into a sequence of 0's and 1's of length less than $H(\xi)$.

Information, therefore, behaves like an incompressible fluid!

So far, we have spoken only of coding by means of sequences of zeros or ones. Whatever has been said so far can be generalized without difficulty in sequences where every element can have one of q possible values. In such a case, instead of (8), we get

(9) $$L \geqq \frac{H(\xi)}{\log_2 q},$$

which can be interpreted similarly: a signal which can assume any of q values contains maximum $\log_2 q$ bits of information. If any particular value of ξ is coded on the average with a code word of length L, in which every element can have one of q values, then a code word, on the average, will have at most $L \log_2 q$ bits of information. And if the code can be uniquely decoded, then $L \log_2 q$ cannot be smaller than $H(\xi)$.

At this point, the notion of redundancy comes up. If a text has symbols, all of which can assume any of q values, and if the text contains H bits of information per symbol, then the redundancy of the text is defined as:

$$R = 1 - \frac{H}{\log_2 q}.$$

R is a number between 0 and 1 which tells how much of the text would be dispensable in case of optimal coding. Shannon investigated the redundancy of written English and found it to be approximately 0.5. This means that, with ideal coding, a written English text (using the same 26 English letters) can be reduced by half. The redundancy of other languages is somewhat smaller but still comparable (30–40%). In connection with this, the lecturer pointed out that it would be a mistake to take the redundancy of these languages as a shortcoming: on the contrary, redundancy has a very important function: it lessens the influence of mistakes such as misspelling, typographical errors, etc.) and, in any case, makes it possible to understand a text. If languages were not strongly redundant, then, in a noisy room (like a crowded restaurant), we wouldn't be able to talk to each other. To look at this in another way: redundancy makes language resistant to noise. Our lecturer mentioned that later on when we investigate the central problem of information theory, the transmission of information via a noisy channel, we will get to the so-called error-correcting codes which automatically correct a certain small number of errors. Now, spoken languages can be looked upon as natural error-correcting codes. We can see the 50% redundancy of a text as follows: if we randomly erase half of the letters in that text we can still, uniquely, reconstruct it from the remaining letters. How much of it can be reconstructed depends, in addition to the language, on the kind of text. For example, the language of the newspapers has a larger redundancy than that of novels, or more particularly of poems, because newspapermen like to use stereotyped expressions, while poets prefer original and unusual adjectives and expressions. As an example, we looked into today's newspaper and the lecturer arbitrarily chose a sentence and erased 50% of it. Wherever he erased a letter, he put a period instead, while a dash stood for the space between two words. What he wrote on the blackboard was the following:

.DMIN.STR.T...–O..IC.A.S–D.–.O.–..CEPT–.H.–

F.R.CA.T.–.F–.–R..ES.I..

We were able to fill in the missing letters without any difficulty. (It was of course helpful that we had already seen versions of this sentence, almost word for word, in previous newspapers.)

What really fascinated me in today's lecture was the point about information behaving like an incompressible fluid. The remarkable thing is that information, which is certainly a matter-like substance, has characteristics similar to matter. What I was thinking was that, although information itself is not matter, it can exist only as connected to matter. Only matter or energy (for example, electromagnetic waves) which in this context, in contrast to information, I will include in the notion of matter, can carry information. A text of letters can be preserved only if it is written or printed on paper, or carved into stone. A voice comes forth in the form of the movements of air-molecules or on records, tapes etc. The transmission of information between two points can happen via cable, with the help of electric current or radio waves. It would be very useful to know in what form our thoughts and memories exist in our brains and how they are recorded. Although we don't know these processes in detail, it is clear that they can happen only in connection with matter, brain matter, that is, the chemical and electrical events of the brain. So it follows that the speed of transmission of information cannot be arbitrarily great, it certainly cannot exceed the speed of light. I thought a lot as well about the notion of redundancy. When I fill in the absent letters in a text, I use the context — but we have learned that the context of information cannot be and is not a subject of information theory. Therefore, the method of investigating the redundancy of a text by erasing and reconstruction is not appropriate. By this method, we would get a correct estimation of the real redundancy only if the reconstruction could be done by a computer. In that case, the meaning of the text wouldn't be a factor because a computer wouldn't understand it and could reconstruct it only by means of a dictionary and grammatical rules. If, for example, the computer reads

TH. .OG B.RKS–.HE C.RA.AN PROGR..S.S,

it will realize that, in the first word, the missing letter is E, and in the third, it is A, since no other letters will produce intelligible words. What about the second word? Here, the computer cannot decide whether to insert a D or an L since judging by grammar alone, the sentence "The log barks" is as correct as "The dog barks", although it is meaningless. The computer cannot determine that a sentence is nonsensical. This example shows well the difficulties to be encountered in programming computers to translate. Finally, I thought about how one might construct a variable-length primitive prefix code (or the appropriate code tree) of less than average code length for a set of messages with a given probability distribution. For $N=2$, the problem is trivial. For $N=3$, the only possible primitive prefix code is the one with a code tree like this:

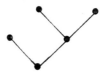

What has to be decided is which of the p_1, p_2, p_3 probabilities should be assigned to the point which is one unit away from the root. It can easily be seen that it must be the largest of the three numbers, and that the two smaller ones should be assigned to the point which is 2 units away. Starting with this, I figured out how to construct, in general, a code of minimal average word length, or the appropriate code tree. The procedure is a recursive one. Let's assume that we know for N code words how to construct the tree. Assume now that we are given $N+1$ words with their probabilities arranged in decreasing order, $p_1 \geq p_2 \geq ... \geq p_{N-1} \geq p_N \geq p_{N+1}$. Let's modify the distribution by deleting the two smallest, p_N and p_{N+1} and by adding to $p_1, ..., p_{N-1}$ the number $p_N^* = p_N + p_{N+1}$. Now we have a probability distribution of N elements and, as assumed previously, we know how to construct a primitive code of minimal average word length or its code tree, having numbers $p_1, ..., p_{N-1}, p_N^*$ assigned to its N terminal nodes. On this tree, let's branch out two new branches from the node with p_N^*, and put the numbers p_N and p_{N+1} at the two terminal nodes. Thus we have arrived at the prefix code of minimal average word length for messages with a $(p_1, ..., p_N, p_{N+1})$ distribution.

An example can make this process crystal clear. Let $N=5$ and the probabilities of the messages be the following:

$$p_1 = \frac{1}{3}, \quad p_2 = \frac{1}{5}, \quad p_3 = \frac{1}{5}, \quad p_4 = \frac{1}{6}, \quad p_5 = \frac{1}{10}.$$

Replacing the two smallest numbers by their sum will result in the distribution $\frac{1}{3}, \frac{4}{15}, \frac{1}{5}, \frac{1}{5}$. Again, combining the two smallest numbers, we get the distribution $\frac{2}{5}, \frac{1}{3}, \frac{4}{15}$. For three-membered distributions we already know what to do, so we end up with the following code tree:

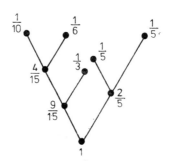

The average word length is $\frac{34}{15} = 2.266...$, while $\sum_{k=1}^{5} p_k \log_2 \frac{1}{p_k} = 2.220...$.

(In the figure, the number at each branching-point of the code tree is the sum of those terminal node numbers which can be reached from that particular branching-point.)

I thought out the proof of this in the general case and it's not at all hard. But I won't set it down here, because since I solved the problem (and I was very happy when I did) I found the same method in one of the books the lecturer gave me. It is called the Huffman code.

At first, I was disappointed that this was already known because I really thought that I had discovered something new. Later, I realized that this was predictable: since we are very much at the introductory level of information theory, it is quite impossible that we should be encountering any unsolved problems. But I'm not sorry about the time I spent solving this problem because it cleared up several matters for me relating to codes and code trees and because it enabled me to understand fully what we had learned. I have read somewhere that you really understand only what you figure out for yourself, like a flower which can only use the water absorbed through its own roots.

Today I spoke to the lecturer and told him how I had "rediscovered" the Huffman-code. He consoled me by pointing out that the value of my efforts is not lessened because I wasn't the first. He also told me about an actual unsolved problem related to the Huffman code: how should the construction be modified if the length of the codewords cannot exceed a given limit? I will think about this problem. Right now, I don't even see why it's such a difficult problem, but obviously it cannot be easy, since if it were, it would have been solved a long time ago. The existence of this problem shows that what I wrote before is not true: even if we are just at the beginning of information theory, there are already, on our current level of knowledge, unsolved, open problems!

Fifth lecture

Today we started to hear about the central problem in information theory, the problem of information transmission through a noisy channel. The professor started with the block diagram set out below:

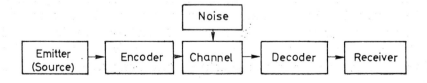

This figure shows the process of information transmission via channel: the emitter, which can be human or machine and which can also be called the information source outputs a certain meassage which is coded by an encoder before transmission. The aim of coding is twofold: to change the message into a form suitable for transmission and to make it "noise-resistant". By the channel we mean the whole physical path between the coder and decoder. In the case of a telegram, the channel is a wire, while in the case of a message sent from a spacecraft, the channel is the whole universe. Because of "noise", information passing through a channel will be randomly distorted or modified. "Noise" means any kind of distorting influence which is random in its effect. For example, if I send a message to a person via a third person, he is the channel and his inaccuracy or lack of attention is the noise source. In the case of a radio or TV in a spacecraft, the channel is the Earth's atmosphere and possible sources of noise are electrical events such as atmospheric electricity, messages from other transmitters on neighbouring wavelengths and electrical appliances such as elevators operating near the transmitter or receiver.

The decoder has a bigger role in case of noisy channel than in a noiseless situation. In this latter situation, decoding is simply a unique translation, as in coding, only backwards. In the case of a noisy channel, the arriving signal is distorted and the decoder must, as best it can, figure out what the undistorted message was and then decode it. The reconstruction of a noise-distorted signal, in general, can be done in many ways; a choice must be made among the several possibilities. Generally, it is assumed that the principle of the decoder is to choose from the possible signals the *most probable one*. These are the *ideal decoders*. To be able to calculate the probabilities of different reconstructions, the decoder needs to know the probabilities of individual code words, in other words, the statistical laws applying to the information source and the probabilities of the different types of distortion introduced by the channel, i.e., the statistical laws of the channel. These can be assumed to be known if the information transmission has been going on regularly for quite some time under basically unchanged circumstances. The theory of information transmission in noisy channels relates to these cases.

There are some cases where the transmitter can have some knowledge about what has reached the receiver. In such cases, we talk about feedback. To this point, we have dealt only with channels into which signals are emitted one after the other, so that each signal can have only finitely many values. Let these values be $a_1, a_2, ..., a_q$. Assume that the noise does or does not distort the signals independently of each other, and that a_i can be distorted only in such a way as to "turn" it into a_j ($i \neq j$). Thus, we are assuming that the receiver messages are the same as the source messages. Denote by ξ a coded signal and η the possibly distorted version of it which reaches the decoder. If ξ is not distorted, then $\eta=\xi$; otherwise $\eta\neq\xi$.

Obviously, $I(\xi,\eta)$ amount of information regulates how well the channel can be used: we are interested in exactly how much information η, the signal received can give about ξ, the transmitted one. Of course, $I(\xi,\eta)$ depends on the distribution of ξ. Let's consider the maximum $I(\xi,\eta)$ for all possible distributions of ξ. This is called *channel capacity c:*

$$c = \max_{\xi} I(\xi,\eta).$$

At a glance, it is evident that c gives only a theoretical limit for the amount of information which can be transmitted through the channel by one signal. This can be proved precisely. Assume that time τ is needed to transmit one signal. Then the speed of information transmission through the channel (i.e. transmitted information per unit time) cannot exceed the limit $\frac{c}{\tau}$. If the information source outputs more information per unit time, it cannot be transmitted through the channel even with the most suitable coding. In such a case, the transmitter must decrease the speed of transmission in some way.

It can be proven that, if we prescribe any velocity which is less than the critical $\frac{c}{\tau}$, it will be possible to transmit information at velocity v through the channel so that it can be reconstructed with an arbitrarily high probability in form close to that which was transmitted. The so-called coding theorems, which will be discussed at the next lecture, prove this fact with different assumptions. These theorems state only the theoretical possibility of information transmission at the specified velocity. They say nothing about the construction of the code. There is another chapter of information theory which deals with such construction: the theory of error-correcting and error-checking codes, which uses algebraic tools (linear algebra, group theory, Galois theory, and finite geometry, as well as the results of modern combinatorics). All of this will come up in future lectures. In this lecture, we worked out the following simple example: Let the input/output signals of the channel be 0 and 1. Assume that in the channel, because of noise, the input signal 1 will become an output signal 0 with probability p (this kind of channel is called a symmetric binary channel). Now

$$P(\eta = 0|\xi = 1) = p, \qquad P(\eta = 1|\xi = 1) = 1-p,$$

$$P(\eta = 0|\eta = 0) = 1-p, \quad P(\eta = 0|\xi = 1) = p.$$

If $P(\xi=0)=w$ and $P(\xi=1)1-w$ then, introducing the notation $q=w(1-p)+(1-w)p$, we arrive at

$$I(\xi, \eta) = h(q) - h(p).$$

$I(\xi, \eta)$ is maximum if $q=\frac{1}{2}$ and therefore

$$c = \max I(\xi, \eta) = 1 - h(p).$$

With simple arithmetic, it can be shown that $q=\frac{1}{2}$ if and only if $w=\frac{1}{2}$ and $q=P(\eta=0)$. This shows that the capacity of a symmetric binary channel is used maximally if the input signals assume the values 0 or 1 with probability $\frac{1}{2}$, so that the output signals will be 0 or 1 with the same probabilities. But the transmitted information will be only $1-h(p)$ of a bit. This is quite understandable because, although every output signal contains 1 bit of information, $h(p)$ bit of it is information about the noise and only $(1-h(p))$ of a bit is information about the input signal. Of course, when $p=\frac{1}{2}$, then $1-h(p)=0$ and no information can be transmitted through the channel, because the output signal is now completely independent of the input signal. Citing this example, the lecturer pointed out that if we are not concerned about transmitting at a speed close to the maximum speed allowed by the channel capacity, in the case of transmission through noisy channel, but, rather, about transmitting information reliably in spite of the noise, we can do so quite easily if every signal is emitted not once, but several times. For example, if every signal is repeated $(2s+1)$ times where $s \geq 1$, then among the $(2s+1)$ signals arriving, there will be a certain number of ones and the rest, $2s+1-r$, will be zeroes.

Now let's do the decoding using the principle of majority: if $r \geq s+1$, then the signal repeated $2s+1$ times will be taken as 1, while if $r \leq s$, it will be taken to be 0. (The reason for transmitting the signal an odd number of times is to avoid getting a 0 or 1 an equal number of times.) Even in this way, it is possible to be mistaken when decoding, but the probability of this can be made as small as we want, if $0<p<\frac{1}{2}$, by making the value of s large enough. The larger s is, the lower will be the speed of information transmission. This coding method is quite primitive; with more complex but also more practical processes, the same result can be obtained (i.e., ensuring an error probability which is arbitrarily small) without a drastic reduction in the speed of transmission.

I tried, when thinking about what I heard today, to make a connection between information transmission through a noisy channel and our game. I made up the following version, which I called "Bar-kochba with lies". Assume that the number of questions which can be asked to figure out the "something" being thought of is fixed and the one who answers is allowed to lie a certain number of times. The questioner, of course, doesn't know which answer is true and which is not. Moreover, the one answering is not required to lie as many times as is allowable.

For example, when only two things can be thought of, and only one lie is allowed, then 3 questions are needed. If the two things one can think of are the numbers 0 and 1, then I will ask three times if it's the 0. If I get a "yes" answer at least 2 times, I can be sure that the other player did in fact have the 0 in mind, since he could lie only once, and therefore when he answered "yes" twice, he could not have been lying, because it would have meant that he had lied twice). Similarly, if I get two "no" answers, 1 is surely the number. If there are four things to choose from and one lie is allowed, then five questions are needed. If two or more lies are allowed, then the calculation of the minimum number of questions is quite complicated. And I think that the game "with lies", when more than one lie is allowed, becomes too complicated for a game as such, but is pretty good at helping one to understand the difficulties encountered in the transmission of information via a noisy channel. It does seem to be a very profound problem: I am very curious to hear the proofs of the coding theorems.

Recently, I read an article somewhere about how we would be able to establish communication with intelligent creatures who also have a high level of technology (if they exist somewhere on another planet in the Milky Way). From what I heard today in the lecture, I would say that this is an information theoretical problem, too. If, for example, these creatures discover us and start to send messages with radiowaves, their messages will reach us in quite a distorted form, mixed up with all the "noise" originating in space. To decode in this case is much harder than in the case of the noisy channel we dealt with in the lecture because in this situation we cannot even be sure whether there really is a message or whether the signal received from space is only random cosmic noise. Moreover, even if these signals do contain a message, we don't have the foggiest idea of what the original signal might have been like. So we don't know what are the input messages we have to decode. This problem is more difficult than the usual problem of information transmission through a noisy channel. Our only hope is that these supposedly existing intelligent creatures are smart enough, not only to realize that there are intelligent creatures on Earth to whom it is possible, technically, to transmit information from such a tremendous distance; but also to transmit it in such a way that we can recognize it as a meaningful message, in spite of the cosmic noise, and can then decipher it. If they are incapable of that much, then we probably won't be losing too much by not being able to make any contact with them.

Yesterday I went to see the movie, *Fahrenheit 451°*. It seems I am completely engrossed in information theory and I try to connect everything with it. The movie made me think how interesting it is that a human being can only memorize a book word by word with enormous difficulty and can only keep it in his memory with an almost heroic effort. On the other hand, educated people have read several hundred — some of them several thousand — books and retain a more or less clear picture of each one without any effort, and those who read intelligently remember the essence of each book. It seems that information isn't stored in the brain in the same way as it is in a book or a computer. It occurred to me suddenly that one characteristic of the human memory which is in contrast to the memory of computers, is that it stores the content of information in some way, i.e., it does exactly what the machine is unable to do. On the other hand the brain is weak in an area in which the machine is at its best: the remembering of long signal series regardless of their content with complete accuracy.

By the way, if we substitute for the total, word by word memorization of a book with a memorization more appropriate to the human brain — that is, relating to its content, or message — then every teacher is a walking book, but one which hasn't been printed. That should be the main goal of education at a university, namely, that the professors should teach only what is missing from the textbooks and exists only in their minds. There are some of our professors who do exactly that, but there are others who read aloud or recite by heart what is in the books and then are surprised that most of the students do not attend the second lecture and that whoever does, does so only to sleep at the back. This year, a visiting foreign mathematician, in his lecture given before the Bolyai Society, after starting a particular theorem said: "and now I will tell you what is the yoga of this theorem". He meant several things by this: that he would try to put into words why the theorem was interesting, what its essence was, which step causes real difficulty in the proof, what kind of trick can be used to overcome that difficulty, how one can think of a theorem like this, etc. In other words, he wanted to talk about those things which cannot be set out in the usual jargon of textbooks, monographs, and essays, since one cannot even give a mathematically precise formulation of them. But an answer reflecting the essence of a problem — the yoga of it, more or less — can be extraordinarily interesting, especially for a beginner. The "book", which we carry in us and whose contents, let us hope, we will some day pass on to the next generation, is shaped by and composed of these very remarks.

Now I will stop writing for a while because I want to think about the talk I have to give next week on the parallel between energy and information.

Preparation for my talk

It seemed at first much more of a philosophical question than a mathematical one, so I wondered if philosophers might already know something about it. I have an enthusiastic, bright philosopher friend and I asked him. He certainly got very excited since energy and information both play important roles in our lives. Although he couldn't tell me anything right away, he promised to do some research on the question.

This, as he told me later, did not prove to be a very fruitful exercise. First, he had looked through the Marxist classics, and then turned to Hegel but to no avail at all. He felt a little offended when I remarked that, to solve new problems, one should not so much look up old quotations, as employ new thinking.

In the meantime, I had realized that, wherever energy has a role, information transmission gets into the picture, too, and from that point on, the analogies become self-evident.

Let's consider the seemingly unrelated area of history. The history of humanity can jokingly be called the history of energy. The importance of the discovery of the energy of fire was tremendous. But to keep the fire burning, to supply it with fuel requires an organized human group. Consequently there arose need for a method to convey information: for hand signs, for speech. Therefore, the discovery of fire can be linked (although not very firmly) to the development of one form of information transmission. But let's go further. Human beings tamed the primitive energies of nature, such as the wind; and wild creatures as well, from which sailing and farming came into existence. Transportation, and the production of goods made it possible for some people to become rich and pay skilled workers to develop new methods of information transmission, namely, the arts — painting and sculpture. The wheel which transforms the energy of a back-and-forth leg action into rotary motion, gave a great impetus to the potter's craft and the decoration of clay pots said many things to the people of that time (and of this).

We can draw a parallel between the discovery of the wheel, the lever, and of simple machines in general and the development of writing, and also between the discovery of the steam-engine and the invention of printing. And, to cite at least one obvious example: it was the use of electricity which made possible the existence of probably the most important information-transmitting machines of today: the telephone, radio and TV.

I wouldn't dare suggest these parallels to my historian friends; they would point out too many flaws in them, by showing either that my logic is faulty or that the time connections are impossible. Still, I think that there is something in them and even if there isn't, what is clear is that the evolution of humanity is closely connected not only to the development of different kinds of energy and energy transmission but also to the development of information transmission. The two greatest recent inventions, atomic energy and computers do not seem to have anything to do with each other. But surely it is clear to everyone that it cannot be purely coincidental that a new energy source and a new method of information processing were discovered at almost the same time.

The level of economic development in a country can be linked almost directly to the amount of energy used. But it can be characterized almost as well by the amount of information in circulation, in which we should include everything, from the daily news to information on the economy and business.

I wasn't particularly surprised to find analogies in biology as well. Every living being is capable of converting energy into energy, or matter (food) into energy. Of course, the higher each species is on the evolutionary scale the more skillful it is and the more complex its functions in this area.

The information conversions carried out by living beings are probably even more complex. When the sensory part of every simple animal comes into contact with food, it is capable of reporting this to its motor system which can then fill its mouth with the food. Light (an image) coming to the eye of an animal will travel via many small neurons to its brain, interestingly enough, with the help of chemical impulses. Then the image will be processed there. As a further example, the conditioned reflex of Pavlov is a very complicated information processing method. But I don't want to write about the human brain and how intricately it can transform and process information because if some creature from Mars finds my diary, he may think us humans show-offs (probably not without reason).

Another example from biology: in the morning, after getting out of bed, one of the first things I want to do is eat, to have something to convert into energy (but, of course, I don't think this over every day, I just feel hungry). At the same time, I start to read the newspaper, as a result of which I sometimes try to put the bread into my nose instead of my mouth. So, simply put,

I'm hungry for information, for something to gnaw on during the day, so I can be annoyed at whatever it was that was badly written in that day's paper. Not only living systems, but any other system (Party, Mathematical Society, factory, Army, anthill) to be able to operate, necessarily requires the circulation not only of energy (matter, money) but also of information among its members.

I think I have found more than enough analogies. If I mention them all during my talk everybody will be bored stiff.

Let's go deeper. What about the matter-like nature of information? Well, if energy is matter, then information must somehow be spirit-natured.

But how can this be so? No, no, no ... information is matter because it can only be transmitted by means of matter (or energy), such as symbols on paper or electrical or chemical impulses.

Later, I recalled that there are certain telepathic, or more scientifically speaking, parapsychological phenomena where the transmission of information is accomplished without matter or energy. Yet, at present, there is not much we know about these phenomena, not even whether they really occur and if so, how. Probably by some form of matter or energy, not yet known to us, in which case even this is not a counterexample to my thought. Fine, but the same information can be transmitted in different ways. For example, the write the same things in the newspapers as they say on the radio. Therefore information must be independent of the matter which transmits it. Now, at last, it is clear. This is the case with energy, too. It is not important in what form we get it, what is important is the amount we receive. So the analogy is essentially complete even from this point of view. I wrote "essentially" because if we prescind the medium of energy, what remains is just a number, its quantity. Doing the same thing with information, we still end up with the information in its entirety, although information theory at present is concerned only with quantity and not quality.

After this debate, I made an agreement with myself to look for an analogy in some more serious relationship.

We know that during the transformation of energy, no energy is lost. Some time ago, I was thinking that when we code a message, what we are really doing is information conversion, like the conversion of the potential energy of water into electrical energy. Is it in any sense true that during conversion, the information doesn't increase or decrease; and if so, in what sense?

One of the papers given me by the professor deals with exactly this question.

If a signal contains $H(\xi)$ information and by coding there will correspond to a signal, on the average, an $L = \sum p_k l_k$ signal, then the principle of information conservation should mean that the L signal of the coded series contains $H(\xi)$ information, i.e., in one of them there will be

(10) $$H = \frac{H(\xi)}{L} \quad \text{information.}$$

This formula (if it is really true) reminds me of the formula describing the energy conversion of a transformer. If there is a secondary coil of L turns for every one turn of the primary coil, then the energy per one volt on the primary side is equal to L volts of the secondary, i.e.,

(11) $$\frac{E_1}{V_1} = \frac{LE_2}{V_2},$$

where E_1 and E_2 denote the energy of the primary and secondary sides respectively and V_1 and V_2 the respective potentials.

Since 1 volt is transformed into L volts, $V_2 = LV_1$ and $E_1 = E_2$ follows from (11). In the case of a transformer then, (11) (which corresponds to (10)) does reflect the conservation of energy.

By analogy to $\frac{E}{V} = I$, $H(\xi)$ and H can be called information current density.

The professor had already referred to (10) in his fourth lecture. In a coded sequence, for every place, there can be q different symbols, which limits H to being at most $\log_2 q$. Then it follows from (10) that $\log_2 q \geq \frac{H(\xi)}{L}$ which is equivalent to (9). Moreover, this shows that we cannot get equivalence in (9) because H cannot be made equal to $\log_2 q$.

But the problem is that H is still not defined. It cannot, as I will show, be defined the same way as $H(\xi)$. If ξ can assume two values x_1 and x_2 with probabilities p_1 and p_2 and these are coded 00 and 01, the first symbol of the code will surely be 0, while the second will be 0 or 1 with probabilities p_1 and p_2, respectively. Therefore, the entropy of the first symbol ($=0$) is different from that of the second. In other words, the entropy per symbol H cannot be defined as the entropy of a symbol as it was in the definition of $H(\xi)$. Next, let's think about taking the entropy of the first n symbols and dividing it by n. In this case, we must also be careful, because the entropy of the first n symbols is not the sum of their individual entropies since the code symbols are not independent. If, for example, the code of x_1 and x_2 are 00 and 11, respectively, then the entropy of the first two symbols is

$$p_1 \log \frac{1}{p_1} + p_2 \log \frac{1}{p_2},$$

although their individual entropy also amounts to this. Naturally, this definition is still no good, because it depends on n. Let's assume that the quotient of the entropy of the first n signals and n approaches a number, defined to be equal to H, as n approaches infinity.

With this definition, (10) can be proven if the coding can be uniquely decoded as the paper suggested. However, I won't discuss it in my talk because the proof is so lengthy. Among the articles, I found another which proves (10) in a more general case (for example, when the individual signals are of finite length in time). It follows from (10) that $L \geq \dfrac{H(\xi)}{\max H}$. Finally, my earlier philosophical considerations led me to solve a problem which is important in practice for a very general case, i.e., what the minimum of the average code length is. All of this shows that the search for analogies (seemingly for their own sake) may not be such an extravagance.

But (10) has a flaw, namely, that it is true only if the codewords can be uniquely decoded. If the code is such that every x_i is coded by 0, then $H=0$, so (10) is not true in general. In the case of energy conversion, one can say that not all of the energy arrives where it is supposed to, but some of it changes into, for example, heat energy. When coding is bad, information will not be lost. It will remain in the original series which had been coded. Thus in this case, the situation is even worse, because if we also count the original information, then adding to it the entropy of the coded series, the total information must generally be more than it was originally. Therefore the parallel doesn't really work in the case of coding which is not uniquely decodable.

During energy conversion, a little bit of the energy (sometimes not so little) always side-steps. In a transformer, it changes into heat-energy and in other cases friction develops. In no case can the entire amount of energy be transmitted, unless it is the case of uniquely decodable coding. This may be so, but after coding, the information would have to be transmitted through an ideal, noiseless channel, which doesn't exist in practice. The probability of a mistake can be very small but never zero. In addition, something like (10) can be proven for such channels. Decode the signal series received from the channel and consider its entropy. The entropy per signal in this series is almost equal to the entropy of the original series. More accurately, if the maximal error probability approaches zero, then the entropy of the decoded series approaches the entropy of the original series.

I didn't find this observation in the papers I was given but I was able to prove it quite easily. If the noise (which can be looked upon as friction in the channel) is not great, then only a small amount of information will be lost, and the less the noise, the less will be the lost information. That the amount lost is equal to the noise is not true, although I'm sure it must be true in some form. I firmly believe that it would be a most significant discovery to find such a relation and to prove it. I don't understand how some of the authors of these papers could have failed to formulate and mention this problem. I guess they are working on it a lot themselves and don't want others to solve it before they do. That is certainly not the right approach. I will definitely talk about this problem and maybe someone with a good idea will be able to advance the discussion of the topic.

The principle of energy conversion makes everyone think of perpetual motion right away. What about perpetual motion in information? In a noiseless channel, i.e., simple coding, there is no information loss, so one can expect to maintain continuous information circulation, just as in the case of frictionless mechanical motion. Imagine, for example, that we have constructed the Huffman-code of the Hungarian alphabet. Let's take a sentence and code it letter by letter with Huffman code, then code this series of 0's and 1's using Morse code (so that we decode it in our imagination) and lastly decode it again so that we end up with the original text. Even if this process is repeated infinitely, the information will remain the same, none will be lost.

This cannot happen with channels in practice. Consider the case of gossip which, as the saying goes, often travels in circles. If we start circulating a piece of gossip, it will be very much distorted when it gets back the first time and sending a new version on its way will result in the reception of an even more distorted one etc., until it has nothing to do with the original.

My conclusion is that in the case of coding and channels which can be easily modelled mathematically, the parallel between information and energy is quite good. What about the case of those difficult events which occur in nature, is it true there too? In crystallization, it looks as though information is born. But, in reality, the structure of the electrons in the atoms has always contained this information; these inherently existing characteristics cause the grid-like arrangement of atoms.

It is curious that this information actually resides in the atoms. Here, then, is another point which disturbs the analogy between information and energy. In energy conversion, the origin of one form is the result of its destruction as another form. In coding, the information of the message to be coded is preserved.

Somehow this analogy of crystallization is not at all clear to me. The origin and evolution of life, which happened some billions of years ago, is quite similar. Can it be explained by saying that the information thus originated had been in the atoms? Ah, I've really gone a little too far if I even want to solve the problem of the origin of life with information theory. Anyway, I'm too sleepy and tomorrow I have to give this talk, so it will be better if I go to bed. I really am nervous. If only I had had a little more time to think over this concluding material a little bit better!

My talk, I think, was very good. We argued a lot. I didn't even have time for the problems of crystallization and the origin of life. At the end, the professor congratulated me. He found my remarks on the principle of information conservation of noisy channels especially interesting. He also emphasized that it's not good to keep quiet about half-formed conclusions, or hunches.

I am eagerly awaiting the future lectures. The professor doesn't look too well.

I hope it's nothing serious.

Games of chance and probability theory

INTRODUCTION

The aim of this article is to introduce by the use of understandable and interesting problems in certain games of chance, especially card games, some concepts and methods of probability theory*. Although I have tried to give a full description of the rules of the games, I have done my best to phrase the examples so that they can be understood by those who are not familiar with the games involved. Of course, those who play bridge will get more out of the problems concerning bridge than those who don't.

One might ask if it is worthwhile to consider card games and games of chance in general from a scientific viewpoint. My opinion is an unconditional "yes". It is worthwhile not only because it helps us to understand combinatorics and probability theory, but also because these problems are interesting in considering the history of science. For example, problems provided by games of chance played quite an important role in the formation of probability theory. Finally, it is also worthwhile because the knowledge gained from the mathematical investigation of games of chance fostered the development of many new ideas in modern science and technology. The concept of shuffling is a good example. It relates not only to the mixing methods used in chemical technology but also to the fundamental notions of thermodynamics.

ON THE SHUFFLING OF CARDS

Whenever we consider questions about a distribution of cards in probability theory, we always assume that the deck of cards to be dealt from is "well shuffled". Card-players use this expression often but since they never define it precisely, let's start by seeing what it means.

---

* I cannot make a detailed probability theoretical evaluation of the individual games here but I will provide references with the discussion of each game indicating where the interested reader can find a more detailed description.

From a probability theoretical point of view, a deck of cards is well shuffled if, after shuffling, all possible sequences of the cards – their permutations – have the same probability. In the case of n cards, there are $n!$ possible permutations* and so a deck of cards is well shuffled if the probability of every sequence is $\frac{1}{n!}$. The probability of an arbitrary event A (where event A means a certain card sequence) is $\frac{k}{n!}$, where k means the number of sequences where A is the outcome. For example, if we shuffle a deck of 52 cards, the probability that the uppermost one is an ace is $\frac{1}{13}$, because among the 52! sequences there are 51! which start with a particular ace. (Since if the uppermost card is the ace of hearts, with the 51 cards underneath one can make 51! different sequences). There are four aces, therefore $k = 4 \cdot 51!$ and the desired probability is

$$\frac{k}{52!} = \frac{4 \cdot 51!}{52!} = \frac{4}{52} = \frac{1}{13}.$$

In reality, shuffling is accomplished either by a player or by a machine making the same movement 10 or 20 times. Every movement means the rearrangement of the deck of cards, i.e., the application of a permutation to the sequence of the cards. The permutation of a set of numbers form a group⁺.

The product of two permutations, for example, P and Q, denoted by PQ, means that first we change the random arrangement by P and then carry out

---

* We can put any card in the first place so there are n possible candidates for this place. No matter which we choose, we cannot now put that one in the second place, so the number of cards available for this position is only $n-1$. We can make our choices regarding the first two places in $n \cdot (n-1)$ ways. Continuing in this way, the resulting total of sequences will be $(n \cdot (n-1) \cdot (n-2)...2 \cdot 1)$, since the last card can be chosen in only one way: the one which remains. The short form of the number $n(n-1)(n-2)...2 \cdot 1$ is $n!$. (Gy. Katona)

⁺ To explain the notion of group requires the introduction of the notion of *operation*. Taking two elements of a given set (in a certain order), we assign to them another element of the set (which can be one of the first two). Examples of such an operation are addition or multiplication on the set of integers. In a group, there is only one operation; let us denote it by the multiplication sign. A set with an operation is called a group if there is an element e such that multiplying any element a of the set by e will result in a, and if for each given element a of the set, there is another which multiplied by a results in a product of e. If the operation is the addition on integers, then $e=0$, because adding it to any number will leave the number unchanged and because, given any number, we can find another which, added to the first, will result in a sum of zero (for example, a and $a(-1)$). If the operation is the multiplication among integers, then $e=1$, but, because we cannot find an integer a for 2 such that $2 \cdot a = 1$, this is not a group. (Gy. Katona)

the Q rearrangement on the resulting sequence. For example, if $n=32$ and P is the

$$1\ldots 16, \quad 17\ldots 32,$$

$$17\ldots 32, \quad 1\ldots 16,$$

permutation, meaning that the card which was originally the 17th is now the first, and that the card which was originally the 18th is now in the second place and so on, until the last place is occupied by the card which was originally the 16th (this can be done by lifting the top 16 cards off the deck, placing them on the table and placing the bottom 16 on top of them), and if $Q=P$, i.e., if we carry out the same operation a second time, then $PQ=P^2=I$ is the identity permutation, so that the result of these two operations will be the original sequence.

Now we can construct two plausible mathematical models (i.e., a simplified picture of the real process). The first we will call a deterministic, and the second a stochastic model.

Let's first assume that any shuffle results in the same type of rearrangement of the deck. Let this permutation be denoted by P. Having carried out this kind of shuffling k times* is equivalent to one rearrangement of the starting configuration, the one which corresponds to the P permutation to the power k.

Such a shuffle is unsatisfactory from a theoretical point of view because (in principle) the result can be calculated exactly.

It is not satisfactory in practice either, the smaller the order of the permutation P, (i.e. that smallest positive number r, for which $P^r=I$, where I denotes the identity permutation, in which each card remains in the same place) the more unsatisfactory it becomes. (The order of permutation P in the above mentioned example is $r=2$.) If r is the order of permutation P this means that the permutations $P, P^2, \ldots, P^r$ are all distinct (but any higher power of P coincides with one of these), therefore no matter how many times we repeat a permutation of the order r we will not be able, theoretically, to produce more than r different arrangements. If there were a permutation P which was of the order $n!$, with its repetition, we could produce every possible order. Such a permutation doesn't exist if $n \geq 3$, because the symmetric group of

---

* Doing P one after the other k times will result in the permutation $P \cdot P \ldots P$ (k times); which can be shortened to read P^k. (Gy. Katona)

order n is not cyclic (moreover it is not even Abelian*, while the cyclic ones are). From a purely mathematical viewpoint, the distribution of permutations according to order is an interesting problem Pál Erdős and Pál Turán deal with in a recent paper [1], but since the movements in shuffling are never repeated exactly (not even when the shuffler is a machine) I won't discuss this point in detail.[+]

The other, more realistic model of shuffling is as follows.

Assume that the result of a shuffle depends on chance also and that a given shuffle can result in only permutation with a certain probability. Assume further that the individual shuffles are independent of each other. This means that if we somehow number all possible $n!$ permutations of n and denote by Q_j the one which corresponds to the number j, then each individual shuffle produces permutation Q_j with probability q_j ($j=1, 2, ..., n!$). (Of course, $\sum_{j=1}^{n!} q_j = 1$.) If the i-th shuffle produces permutation Π_i, then Π_i is a random permutation with distribution

$$P(\Pi_i = Q_j) = q_j \quad (j = 1, 2, ..., n!, \ i = 1, 2, ...),$$

i.e., the random permutation Π_i will be equal to Q_j with probability q_j and the distribution of Π_i is independent of $\Pi_1, ..., \Pi_{i-1}$ permutations.

In this case, after the k-th shuffle (if the original configuration of the cards was the $1, 2, ..., n$) we get the permutation

$$\Pi_1, \Pi_2, ..., \Pi_k = \Pi^{(k)}.$$

The $\Pi^{(k)}$ permutations constitute a so-called Markov chain[±]. It is known that if the distribution $\{q_j\}$ is such that for an arbitrary subgroup G of the

---

* Symmetric group of order n: the group of n element permutations.
  Cyclic group: there is an element a in the group such that all other elements of the group are powers of a.
  Abelian group: the product of every multiplication will be unchanged if the order of the factors is changed. (Gy. Katona)

[+] Erdős and Turán prove in [1] that among the $n!$ permutations, most of them have order between the limits $e^{(1/2-\varepsilon)} \log^2 n$ and $e^{(1/2+\varepsilon)} \log^2 n$, where $\varepsilon>0$ is an arbitrarily small number if n is large enough (here $\log n$ means the natural logarithm of n). In the case of the shuffling of cards, this means that, in most of the permutations of the 52 cards, the shuffler need repeat his movements less than 200,000 times to get the cards back to the original order. Compare this to the number 52!, which has 68 digits.

[±] All this means is that if $\pi^{(k-1)}$ is fixed, then $\pi^{(k)}$ does not depend on $\pi^{(k-2)}$ or any of its predecessors), in other words, $\pi^{(k)}$ depends on these previous permutations only through $\pi^{(k-1)}$. (Gy. Katona)

group of all permutations

(2.1)
$$\sum_{Q_j \in G} q_j < 1,$$

(i.e., the distribution is not concentrated in a subgroup), then for a large enough k, the distribution of $\Pi^{(k)}$ will be almost uniform, meaning that any permutation has the same probability of occurring $\left(\text{about } \dfrac{1}{n!}\right)$ after a large number of shuffles. More precisely in this case*

(2.2)
$$\lim_{k \to \infty} P(\Pi^{(k)} = Q_j) = \frac{1}{n!} \quad (j = 1, 2, \ldots, n!).$$

[Condition (2.1) will be satisfied if for every j the $q_j > 0$ relation holds.]

Practically speaking, it follows from the above that if every movement in shuffling is random, the result can be any configuration. If we undertake a large enough number of shuffling movements, then the assumption that "the deck is well shuffled" is valid. We will not deal here with what is meant by a "large enough" number of movements.

Certainly it was worthwhile to consider the process of shuffling in such detail since it is well known that the most frequently used trick of card sharks is insufficient shuffling (see [5] too.)

PROBLEMS ON THE DISTRIBUTION OF CARDS

Let's investigate some simple examples!

A) Poker. Every player is dealt 5 cards of the 52. The following configurations are distinguished in the game:

1. Straight Flush: All 5 cards are of the same suit and in sequence. (The ace can be taken as a 1 or as the card that follows the king, for example, 9, 10, Jack, Queen, King of Hearts.
2. Four of a kind: Four of the five are the same (for example, 4 kings), the fifth can be anything.
3. Full House: Three of the five are of one value and the remaining two are a pair (for example, three 10's and two Kings).
4. Flush: All five cards are of the same suit (for example, all are clubs).

---

* This theorem is a special case of the generalization of the central limit theorem in probability theory for topological groups which gives a condition for convergence in the Haar measure. (See [2] and [3].)

5. Straight: The five cards are in sequence but are not all of the same suit, (for example, 5 of clubs, 6 of clubs, 7 of spades, 8 of hearts, 9 of diamonds).
6. Three of a kind: three cards are the same (for example, three 7's), the other two can be anything (but different from each other, otherwise the hand would be a Full House).
7. Two pair: Two sets of two are the same (for example, two aces and two 6's) and the fifth has any value other than those represented in the pairs.
8. One pair: two cards are the same (for example, two Queens), the three remaining ones can be anything, so long as they differ from each other and from the pair.

The probabilities of these hands can be calculated if we calculate the number of ways in which each of them can possibly occur and divide these numbers by the number of all possible combinations of five cards out of 52. For example, four of a kind can occur in $13 \cdot 48$ ways, while you can pick 5 cards out of 52 in $\binom{52}{5}$ different ways*. The probability of four of a kind is therefore

$$\frac{13 \cdot 48}{\binom{52}{5}} = \frac{13 \cdot 48 \cdot 120}{52 \cdot 51 \cdot 50 \cdot 49 \cdot 48} = \frac{1}{4165} = 0.000240.$$

It may appear that we have calculated in a different way here than before because we did not consider the order of the cards. But this does not influence the final result, because five cards can be arranged into $5! = 120$ configurations and leaving this factor out of the numerator and denominator will not change the value of the ratio. Employing a method similar to the one discussed previously, the probability of four of a kind can be calculated as follows. Assuming that there are four players and that the cards are dealt one to each in turn, the player who is first in order will get the first, fifth, ninth, thirteenth and seventeenth cards of the deck. The number of configurations of the deck where there are five fixed cards in these places is 47! because

---

* $\binom{52}{5}$ is the shorthand notation for $\frac{52!}{5!\,47!} = \frac{52 \cdot 51 \cdot 50 \cdot 49 \cdot 48}{5!}$. It is easy to prove the last expression. If we consider the order in which the five cards were dealt, then the result is $52 \cdot 51 \cdot 50 \cdot 49 \cdot 48$, since any card can be the first and any card but the one dealt first can be the second, so that there are only 51 possible choices for the second position, etc. But in this way we will count each set of 5 cards 5! times, i.e., as many times as there are possible configurations of five cards. That is the reason for dividing by 5!

In the product $13 \cdot 48$, the number 13 signifies the 13 possible combinations in which we can choose the same card from each of the four suits, while the fifth card can be any of the remaining 48. (Gy. Katona)

that is the number of ways in which the other 47 cards can be arranged in the 47 remaining places. Therefore the probability that a player will be dealt four of a kind is:

$$\frac{13 \cdot 48 \cdot 5! \, 47!}{52!} = \frac{13 \cdot 48}{\binom{52}{5}} = \frac{1}{4165},$$

which is the same result as before. The probabilities of different hands are summarized in the following table (to 6 decimal places)

Straight Flush	0.000 014
Four of a kind	0.000 240
Full House	0.001 385
Flush	0.001 967
Straight	0.003 532
Three of a kind	0.021 055
Two pair	0.047 373
One pair	0.422 570
No combination	0.501 864
Total:	1.000 000

In poker, the more a hand is worth, the smaller is its probability. (The order of the probabilities in the above table can change if we don't play with 52 cards or if there are one or more Jokers in the deck.)

According to the rules of poker, the players, after looking at their cards anteing, can discard some of their cards and ask for others. While the probability of not getting any combination in the first deal is somewhat larger than $\frac{1}{2}$, if a player subsequently asks for 5 new cards, the probability that he will still not get any combination is exactly $\frac{1}{2}$. But since the two events are almost independent, the probability that no one will have any combination after the second deal is only approximately $\frac{1}{4}$. If I play poker with three other players, the probability that none of them has even one pair is approx.* $\frac{1}{64}$, which means that if I have only one pair I can be almost sure that at east one of the other three also has a pair, or a better hand. To win with a 1

---

* The events are not completely independent but close to it, so we will not commit a large error if we multiply the probabilities.

pair, therefore, is quite improbable. That one can bluff in poker is another consideration.

We refer the reader to K. Jordan's book [4] which contains many more probability theoretical problems concerning poker.

B) Bridge.* The 52 cards are dealt to four players, who are divided in teams of two. There are two parts to the game: the bidding and the actual play. Many bidding systems are known. In the Culbertson system (see Ref. [5]), the players first evaluate their own cards and decide accordingly what to bid as a Trump suit. The evaluation is carried out by counting the so-called "Tricks" according to the following rules:

Ace (without the king and queen of the same suit): 1 Trick
ace and king (without the queen of the same suit): 2 Tricks
ace and queen (without the king of the same suit): 1.5 Tricks
ace, king and queen of the same suit: 2.5 Tricks
king (without the ace and queen of the same suit): 0.5 Trick
king and queen of the same suit (without the ace): 1 Trick

The value of the hand is determined by the sum of the tricks. For example, if the hand is as follows:

SPADES:	HEARTS:	DIAMONDS:	CLUBS:
ACE QUEEN	KING QUEEN	QUEEN 7	8
10 8 7	4 3 2		

its value is $1.5 + 1 = 2.5$ tricks.

The value of any hand is a number determined by chance, therefore it is a random variable.

Let us now investigate what is the expected value of a hand.+

---

* Bridge is not strictly a game of chance, because the skill of the players is more important than chance. The distribution of the cards, nevertheless, depends on chance so many probability theoretical problems arise here, too.

+ The expected value of a random variable is determined by calculating the weighted sum of all the possible values of the random variable (where the weights are the appropriate probabilities of the particular value). In other words, if the random variable can assume the values $x_1, x_2, ..., x_n$ with probabilities $p_1, p_2, ..., p_n$ respectively, then its expected value is

$$E(\xi) = p_1 x_1 + p_2 x_2 + ... + p_n x_n.$$

The importance of the expected value can be demonstrated by the law of large numbers.

If we observe the values assumed by a random variable in a large enough number of independent trials, then the algebraic average of the assumed values will almost certainly be very close to the expected value of the random variable.

To be able to do so, we would first have to calculate the distribution of the hand, i.e., the probabilities that a certain player has a hand of 13 cards of value 0, 0.5, 1, 1.5, 2, 2.5, 3, 3.5, 4, 4.5, 5, 5.5, 6, 6.5, 7, 7.5, 8, 8.5, 9, 9.5, 10, respectively. (It is easy to see that there are the possible trick values.) Although this is not too difficult to do, we will adopt a simpler method, using a property of the expected value, namely that the expected value of the sum of random variables is equal to the sum of the expected values of these random variables. Let's denote by ε_1, ε_2, ε_3 and ε_4 the values of the hands of the four players. Obviously, if the deck of cards is well shuffled, the expected values of all four hands are the same. The trick value of the first player's hand also consists of four parts:

$$\varepsilon_1 = \varepsilon_{11} + \varepsilon_{12} + \varepsilon_{13} + \varepsilon_{14},$$

where ε_{11}, ε_{12}, ε_{13} and ε_{14} stand for the trick values in spades, hearts, diamonds and clubs, respectively.

Similarly, the values of the hands held by the other players can be written as:

$$\varepsilon_2 = \varepsilon_{21} + \varepsilon_{22} + \varepsilon_{23} + \varepsilon_{24},$$
$$\varepsilon_3 = \varepsilon_{31} + \varepsilon_{32} + \varepsilon_{33} + \varepsilon_{34},$$
$$\varepsilon_4 = \varepsilon_{41} + \varepsilon_{42} + \varepsilon_{43} + \varepsilon_{44}.$$

The expected values of the random variables ε_{ij} ($i=1, 2, 3, 4; j=1, 2, 3, 4$) are the same. If m denotes this common value, i.e.,

$$E(\varepsilon_{ij}) = m \quad (i, j = 1, 2, 3, 4),$$

then, because of the additivity of the expected values,

$$E(\varepsilon_1) = 4m = E(\varepsilon_{11} + \varepsilon_{21} + \varepsilon_{31} + \varepsilon_{41}).$$

Since $\varepsilon_{11} + \varepsilon_{21} + \varepsilon_{31} + \varepsilon_{41}$ is the sum of the trick values of the four hands in spades, we have demonstrated that the expected value of any hand is equal to the expected value of the sum of the spade tricks in the four hands. To calculate this value, we will first have to investigate how the ace, king and queen of spades are distributed among the four players. If all three are in different hands, then $\varepsilon_{11} + \varepsilon_{21} + \varepsilon_{31} + \varepsilon_{41} = 1.5$. If the ace and king are in one hand and the queen in another, or if the ace and queen are together while the king is elsewhere or finally, if the king and queen both happen to have been dealt to one hand, and the ace to another, then $\varepsilon_{11} + \varepsilon_{21} + \varepsilon_{31} + \varepsilon_{41} = 2$. Lastly, if all three are in one hand, then $\varepsilon_{11} + \varepsilon_{21} + \varepsilon_{31} + \varepsilon_{41} = 2.5$.

We can calculate the probability that the ace, king and queen of spades are in different hands. If player A receives the ace of spades, he will be dealt another 12 cards from the 41 and the other three players will get 39. The probability that A will not get the king of spades is therefore $\frac{39}{51}$. If player A and player B receive the ace and king of spades respectively, then the probability that the queen of spades will go to C or D is, reasoning as above, $\frac{26}{50}$.

Thus, the probability in question is:

$$\frac{39}{51} \cdot \frac{26}{50} = \frac{169}{425} = 0.398.$$

Similarly, it can be seen that the probability that two of these cards are in the same hand and the third in another is

$$3 \cdot \frac{12}{51} \cdot \frac{39}{50} = \frac{234}{425} = 0.550.$$

And the probability that the ace, king and queen are in the same hand is

$$\frac{4\binom{13}{3}}{\binom{52}{3}} = \frac{22}{425} = 0.052.$$

The sum of these three should, of course, be equal to one: $0.398+0.550+ +0.052=1$.

The expected value of the value of a player's hand is thus: $1.5 \cdot 0.398+ +2 \cdot 0.550+2.5 \cdot 0.052=0.597+1.100+0.130=1.827 \sim 1.8$. This result substantiates the rule of the Culbertson system that at least 2.5 tricks are needed to start bidding since to have 2 tricks is to have a not better than average hand. On the other hand, if the player's partner has started the bidding, then 1.5 tricks, i.e., a close-to-the-average hand, is enough to respond. The reason for this is that the expected value of the sum of the expected values of the two opponents' hands is 3.6; while with $2.5+1.5=4$, the partner of the player who started the bidding can expect to be stronger together than the other side.

Another instructive question in connection with bridge is as follows: if a player has two aces, what are the probabilities that the other two are in his partner's hand, that only one of them is there, or that he has none of them? Obviously, the other two aces are among the three other hands dealt out of the 39 remaining cards and they can be distributed in $\binom{39}{2}$ ways, of which $\binom{13}{2}$ are valid in relation to the first question. Therefore, the probability that the partner has the other two aces is

$$\frac{\binom{13}{2}}{\binom{39}{2}} = \frac{6}{57}.$$

The probability that only one ace is in the partner's hand is:

$$\frac{13 \cdot 26}{\binom{39}{2}} = \frac{26}{57},$$

while the probability that he has none of them is:

$$\frac{\binom{26}{2}}{\binom{39}{2}} = \frac{25}{57}.$$

The sum of the three probabilities of course, is equal to 1:

$$\frac{6+26+25}{57} = 1.$$

A detailed probability theoretical treatise on bridge which is understandable to laymen can be found in a book by E. Borel and A. Chéron [Ref. 6.].

GAME STRATEGIES

In this section, we will restrict ourselves to the following simplified game of chance. There is one player (let's call him Peter) against the Bank. The game consists of a series of runs. At every run Peter can decide how much he wants to risk and this amount is called his *bet*. He must put his bet on the table, therefore he can never bet more than the money he actually has with him. Then they carry out a random experiment which has two possible outcomes, the events A and $\bar{A}$, with probability p and q (where $p+q=1$, $0<p<1$) respectively. If the outcome happens to be A, Peter can keep his bet and the Bank pays him the amount of his bet. If the event $\bar{A}$ is the result, the Bank collects Peter's bet.

Coin tossing is such a game with $p=\frac{1}{2}$ (if we assume regular coins). Roulette is another example, if we assume that Peter will always choose a red number. In that case, since there are 18 red and 18 black positive numbers, and a zero (the bank also wins if the outcome is zero), on the revolving wheel, $p = \frac{18}{37}$.

It is well known that if $p \leq \frac{1}{2}$, there is no system to ensure that Peter will win. Let us restrict ourselves to the $p = \frac{1}{2}$ case. Let $\varepsilon_k = +1$ if Peter wins on the kth spin (i.e., if the outcome is event A) and $\varepsilon_k = -1$ if he loses (so that $\bar{A}$ occurred). Let S_k denote Peter's bet on the kth spin, S_k may depend on $\varepsilon_1, \varepsilon_2, ..., \varepsilon_{k-1}$: $S_k = S_k(\varepsilon_1, ..., \varepsilon_{k-1})$. Each S_k always assumes non-negative values. $S_k = 0$ means that Peter does not bet on the kth spin. $S_n \neq 0$ and $S_j = 0$, if $j > n$, means that Peter stops playing after the nth spin.

If Peter sits down with N dollars, we understand as one of Peter's possible strategies an arbitrary series of the non-negative functions $S_k(\varepsilon_1, ..., \varepsilon_{k-1})$ ($k = 1, 2, ...$) (where S_1, is constant and the variables ε_i can assume either ± 1 values) provided that

$$N + \sum_{k=1}^{n} \varepsilon_k S_k(\varepsilon_1, ..., \varepsilon_{k-1}) \geq 0 \quad \text{for} \quad (n = 1, 2, ...).$$

Let $\xi_0 = N$.

The sum $\xi_n = N + \sum_{k=1}^{n} \varepsilon_k S_k(\varepsilon_1, ..., \varepsilon_{k-1})$ ($n = 1, 2, ...$) indicates how much money Peter has after the nth spin. The random variables ξ_n ($n = 0, 1, ...$) form a so-called *martingale* (Ref. [7]), which means that the expected value of ξ_n given

$$\xi_1, \xi_2, ..., \xi_{n-1}$$

is always equal to ξ_{n-1}.

It can easily be seen in the case $p = \frac{1}{2}$, that $E(\xi_n) = \xi_{n-1}$ ($n = 1, 2, ...$), thus no system will guarantee that Peter wins for sure. It is worthwhile to spend some time with the following faulty "system" — popular among gamblers who don't know probability theory. This system holds that Peter should keep betting 1 dollar until he is 1 dollar ahead for the first time. At that point, he should quit. It does seem that this system ensures that Peter will win 1 dollar, since (with probability 1) sooner or later he will win. In reality, this is not a winning strategy. Obviously this system is not justified according to the rules above because if Peter loses in the first N spins, or if during the first $N + 2M$ games, he loses a total of $N + M$ times and wins M times in such a way that he is never ahead in the meantime, then he cannot continue in the game and (with a positive probability) will lose all his money. In fact, his expected winnings in such a game is 0 and this can be proven as follows. Let

$f_k(N)$ denote the probability that Peter loses all his N dollars instead of winning $k-N$ dollars. If $N \geq 2$, then

$$(4.1) \qquad f_k(N) = \frac{1}{2}f_k(N+1) + \frac{1}{2}f_k(N-1),$$

This event can happen if Peter loses on the first spin and then loses his remaining $(N-1)$ dollars on the next one instead of gaining $(K-N+1)$, or if the wins on the first spin and then loses $(N+1)$ dollars instead of winning $(K-N-1)$.

It can easily be seen* that all possible solutions of the difference equation (4.1) have the form of $f_k(N) = AN + B$. Since $f_k(0) = 1$ (i.e., if Peter has no money he cannot play and therefore he cannot win) and $f_k(k) = 0$ (because if at the beginning of the game Peter had k dollars, he didn't even have to play) therefore $f_k(N) = 1 - \frac{N}{k}$. We are interested in $k = N+1$, so the probability that Peter will lose N dollars before gaining the one is $\frac{1}{N+1}$. Accordingly, the expected value of his winnings using this system is

$$1\left(1 - \frac{1}{N+1}\right) - N \cdot \frac{1}{N+1} = 0.$$

From what has been said so far the reader could well conclude that what probability theory can tell the gambler is that if he is only playing for the winnings (and not for the joy of the game), then it's better for him not to play at all. But that is not the case. If what the gambler asks of the mathematician is to work out a system which guarantees winning, then the wish is an untenable one and a mathematician cannot help. If, on the other hand, the gambler's aim is reasonable, the mathematician can tell him the best way to reach his goal.

Let us first consider the following problem. Peter plays heads or tails. At the start, he has N dollars, and he decides to play until his money amounts

---

* The proof is as follows: if $f_k(N)$ satisfies (4.1), then $g(N) = f_k(N) - \frac{N}{k}(f_k(k) - f_k(0)) - f_k(0)$ satisfies (4.1) as well and $g(0) = g(k) = 0$. Let $\max g(N) = G(N_1)$, then by induction it can be seen that $g(N_1 + j) = G$ ($j = 1, 2, \ldots, k - N_1$). Similarly $g(N_1 - j) = G$ ($j = 1, 2, \ldots, N_1$) therefore $g(N) = 0$, if $N = 0, 1, \ldots, k$ and so $f_k(N) = AN + B$ where $A = \frac{1}{k}(f_k(k) - f_k(0))$ and $B = f_k(0)$.

to $M > N$ or until he loses everything. For which system will the probability of his winning be greatest? Let $w = w(N, M)$ denote the probability that Peter will stop playing with M dollars. Since he cannot lose more than N dollars and since the expected value of his winnings should be 0, therefore $w(M-N) - (1-w)N = wM \le N = 0$ which implies $w \le \frac{N}{M}$. The question is with which system will the probability of winning be $w = \frac{N}{M}$? Let us suppose a "daring" strategy where Peter bets all his money while he has $\le \frac{M}{2}$; puts up $(M-x)$ until $x > \frac{M}{2}$ but $x < M$, i.e., exactly the amount with which if he wins on the next toss, he will end up with the M dollars he was aiming for. For example, if $N = 1$ and $M = 10$, then Peter's play will be as follows: On the first toss he will bet 1 dollar; if he loses, he will have to quit and if he wins he will have 2 dollars. In the latter case, he will bet 2 dollars on the second toss, so that if he loses, he can depart with sorrow and if he wins, he can take all his money (i.e., 4 dollars) on the next toss. As a loser in that round, he will go home emptyhanded but he will have 8 dollars if he wins. Now he will bet only 2 dollars so that if he wins, he will already have his 10 dollars and can quit. Even if he loses, he will still have 6 dollars and will be able to continue playing. By betting 4 dollars on the next toss, he will have 10 if he wins and can go home happy and even if he loses, will still have 2 dollars. He can still play with those 2 dollars — and so on. In this numerical example, the flow of Peter's money can be shown with the following directed graph (see Fig. 1).

There are two edges (of probability $\frac{1}{2}$) leading from every vertex. If p_i is the probability that the player will get from point i to point 10, then the

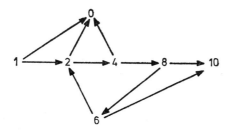

Fig. 1

equations set out below follow:

$$p_8 = \frac{1}{2} + \frac{1}{2} \cdot p_6,$$

$$p_6 = \frac{1}{2} + \frac{1}{2} \cdot p_2,$$

$$p_4 = \frac{1}{2} \cdot p_8,$$

$$p_2 = \frac{1}{2} \cdot p_4,$$

$$p_1 = \frac{1}{2} \cdot p_2.$$

By substitution, this system of linear equations can be solved in the following way (because its determinant is not 0):

$$p_2 = 2p_1, \quad p_4 = 4p_1, \quad p_8 = 8p_1,$$

furthermore,

$$16p_1 - p_6 = 1, \quad p_6 - p_1 = \frac{1}{2},$$

which gives

$$p_1 = \frac{1}{10}$$

and thus

$$p_i = \frac{i}{10} \quad \text{for} \quad i = 1, 2, 4, 6, 8.$$

Therefore, $w(1, 10) = \frac{1}{10}$. Similarly, we can determine $w(N, M)$ when N and M are arbitrary positive numbers such that $N < M$ and $\frac{M}{N}$ is a rational number.

If N and $M > N$ are very large numbers, this method is not as appropriate because of the large number of equations we would end up with. Accordingly, in the general case, we should apply another method. It is true in general if N and M are arbitrary positive numbers (not necessarily integers) and $N < M$, then $w(N, M) = \frac{N}{M}$. We can demonstrate this as follows. Assume that $M = 1$ and $0 < N < 1$, since we can choose M to be our monetary unit. Let $w(N, 1) =$

$= f(N)$ $(0 \leq N \leq 1)$, then the following equation is obviously true:

(4.2) $$f(x) = \begin{cases} \dfrac{1}{2}f(2x) & \text{for } 0 \leq x \leq \dfrac{1}{2}, \\ \dfrac{1}{2} + \dfrac{1}{2}f(2x-1) & \text{for } \dfrac{1}{2} \leq x \leq 1. \end{cases}$$

This equation can be solved as we solved (4.1). Let $g(x) = f(x) - x$, then $g(x)$ satisfies (4.3)

(4.3) $$g(x) = \begin{cases} \dfrac{1}{2}g(2x) & \text{for } 0 \leq x \leq \dfrac{1}{2}, \\ \dfrac{1}{2}g(2x-1) & \text{for } \dfrac{1}{2} \leq x \leq 1. \end{cases}$$

This being so, $g(x)$ is bounded, $-1 \leq g(x) \leq 1$ since $f(x)$ is a probability and thus $0 \leq f(x) \leq 1$. Let $G = \sup_{0 \leq x \leq 1} g(x)$ and x_n be a sequence such that $\lim_{n \to \infty} g(x_n) = G$.

From the (bounded) sequence x_n, a convergent subsequence y_n can be chosen, to yield

$$\lim_{n \to \infty} y_n = y \quad \text{and} \quad \lim_{n \to \infty} g(y_n) = G.$$

If $0 \leq y_n \leq \dfrac{1}{2}$ for infinitely many n, then according to (4.3),

$$G = \lim_{n \to \infty} g(y_n) \leq \frac{1}{2} \limsup_{n \to \infty} g(2y_n) \leq \frac{G}{2},$$

while $\dfrac{1}{2} \leq y_n \leq 1$ for infinitely many n, so

$$G = \lim_{n \to \infty} g(y_n) \leq \frac{1}{2} \limsup_{n \to \infty} g(2y_n - 1) \leq \frac{G}{2},$$

so that $G \leq \dfrac{G}{2}$ in all cases, i.e., $G \leq 0$.

Now let $g = \inf_{0 \leq x \leq 1} g(x)$. Using a similar reasoning, it can be shown that $g \geq 0$, which means $g = G = 0$, i.e., $g(x) \equiv 0$ and therefore $f(x) \equiv x$, which is exactly what we wanted to prove.

It should be noted that in the game of tossing coins each strategy which results in Peter losing all his money with probability one in finitely many runs or winning the desired amount, the probability of winning will be the same as in the "daring" strategy discussed above.

On the other hand, let's examine a game where Peter wins his bet in each case with probability p and loses it with probability $q = 1 - p$, where $0 < p <$

$< \frac{1}{2}$. (Roulette is such a game; if Peter always chooses a red, $p = \frac{18}{37}$, as we have seen.) In this case, it does matter what kind of strategy Peter chooses and the "daring" strategy is indeed optimal. Let's assume again that, at the beginning of the game, he has a sum of money x, where $0 < x < 1$, his goal being to end up with this sum being equal to one. Let $g(x, p)$ denote the probability that Peter will achieve his goal if he uses the "daring" strategy.

As in the case where $p = \frac{1}{2}$ when we arrived at (4.2), now we get

(4.4) $\quad g(x, p) = \begin{cases} pq(2x, p) & \text{for } 0 \le x \le \frac{1}{2}, \\ p + (1-p)g(2x-1, p) & \text{for } \frac{1}{2} \le x \le 1, \end{cases}$

provided that $g(0, p) = 0$ and $g(1, p) = 1$. In the same way that we concluded that the function $f(x)$ satisfying (4.2) must be identical to x, we now conclude that there is a unique solution to (4.4) which satisfies the conditions $g(0, p) = 0$, $g(1, p) = 1$. (This was first proven by G. de Rham [Ref. (10)]). One solution of (4.4) can be constructed in the following way. Let $\xi_1, \xi_2, \ldots$ be independent random variables assuming the values 0 and 1 with probabilities p and $1-p$ respectively.

Let $\eta = \sum_{n=1}^{\infty} \frac{\xi_n}{2^n}$ and let $F_p(x)$ denote the distribution function* of random variable η. Then $F_p(x)$ satisfies the equation

$F_p(x) = \begin{cases} pF_p(2x) & \text{for } 0 \le x \le \frac{1}{2}, \\ p + (1-p) \cdot F_p(2x-1) & \text{for } \frac{1}{2} \le x \le 1, \end{cases}$

and the conditions $F_p(0) = 0$, $F_p(1) = 1$. Thus $F_p(x) = g(x, p)$.

The measure $\mu_p(A)$ in which $\mu_p(I_{a,b}) = F_p(b) - F_p(a)$ for $0 \le a < b \le 1$, where $I_{a,b}$ denotes the interval $a \le x < b$ on the Borel-sets of the interval $(0, 1)$ can be characterized as follows. The measure of the interval $\left(0, \frac{1}{2}\right)$ is p, and the measure of the interval $\left(\frac{1}{2}, 1\right)$ is $1-p$. The measure of the interval

---

* The probability distribution function of a random variable η is $F(x) = P(\eta < x)$, i.e., $F(x)$ gives the probability of η being less than x. (Gy. Katona)

$\left(0, \frac{1}{2}\right)$ is divided in the proportion $p/(1-p)$ between the intervals $\left(0, \frac{1}{4}\right)$ and $\left(\frac{1}{4}, \frac{1}{2}\right)$: a similar proportionality is used for the interval $\left(\frac{1}{2}, 1\right)$, etc. Therefore there will be $\binom{n}{l}$ intervals having measure $p^l(1-p)^{n-l}$ for $l = 0, 1, \ldots, n$ among the $\left(\frac{k}{2^n}, \frac{k+1}{2^n}\right)$ ($k = 0, 1, \ldots, 2^n-1$) intervals. $F_p(x)$ is a strictly monotone increasing function, continuous and singular; its derivative is in almost all cases 0. The measures μ_{p_1} and μ_{p_2} are orthogonal if $p_1 \neq p_2$. Clearly $\mu_{1/2}$ is equivalent to the Lebesgue-measure since

$$F_{1/2}(x) = g\left(x, \frac{1}{2}\right) = x \quad (0 \leq x \leq 1).$$

We will not show here that, in the case of $p < \frac{1}{2}$, it does matter what kind of strategy one employs, and that the "daring" strategy is optimal. Instead, we will give an illustrative example.

Let us assume that Peter sits down to play roulette $\left(p = \frac{18}{37}\right)$ with 25 dollars and that he will use the "daring" strategy to try and win 100 dollars. He will succeed if and only if he wins on the first two spins and the probability of that happening is $p^2 = 0.244\ldots$. Now let us see how he would do if he were to apply a more cautious strategy, always betting only 25 dollars. He would accomplish his goal with a probability of $\dfrac{p^3}{1-2p+2p^2}$ and $\dfrac{p^3}{1-2p+2p^2} < p^2$ if $p < \frac{1}{2}$, because

$$p^2 - \frac{p^3}{1-2p+2p^2} = \frac{p^2(1-p)(1-2p)}{p^2+(1-p)^2} > 0.$$

If $p = \frac{18}{37}$, then $\dfrac{p^3}{1-2p+2p^2} = 0.2301$, indicating that Peter's chance of winning is more than 23.5% using the "daring" strategy, while it is somewhat less in the case of the "cautious".

The book strategy by L. E. Dubbins and L. J. Sarage deals with similar, more general problems (Ref. [8]).

A MATHEMATICIAN'S WAR AGAINST THE CASINOS

To finish up, we will tell of a curious case which shows what the mathematical theory of games can and cannot do. An American mathematician, Edward O. Thorp, who was teaching at a university in Los Angeles went to Las Vegas for a couple of days during the winter holiday. He visited one of the casinos there, played twenty-one (blackjack) — and lost. Annoyed, he began to think about what would be the best strategy to use in playing blackjack (as it used to be played in the Nevada casinos).

Blackjack is played as follows. The dealer (an employee of the casino) deals two cards to each player from a thoroughly shuffled deck of 52 cards. The players don't show their cards to the dealer, but the dealer, who also gets two cards, has to show his first card to the players. The cards have the following values: 10 points for each court card (jack, queen, king), and face value for all other cards (for example, a 7 counts for seven points) except for the ace, which, at the player's option can be valued as either 1 or 11. The winner is the one whose hand value comes closest to 21 without exceeding it.

Each player, after looking at his hand, can ask for as many additional cards as he wishes, but if the total value exceeds 21 he has to show his cards and is out of that round. The dealer can also deal further cards to himself. Bets are made freely between an upper and a lower limit. Every player plays against the dealer. If a player has a better hand than the dealer he wins as much as he bet. If his hand is worse, then he loses his stake. And in the case of equality (for example if both are 21) no money changes hands.

The big advantage for the dealer comes from the fact that the players always have to show their hand when its value exceeds 21. In addition, they lose their bet even if the dealer has more than 21, because he doesn't have to show his hand if everyone else's hand has already exceeded 21.

Thorp realized that the casino set very strict rules in accordance with which their employees have to play*. For example, if the dealer's hand is or exceeds 17 he cannot "request" another card for himself. Thorp thought that the freedom a player has (as opposed to the dealer), in that he does not have to show his first card and can decide the amount of his bet, in theory, makes it possible to work out a winning strategy. His most important observation was that at that time in the casinos of Nevada, in order to save time, the dealer did not shuffle the entire deck of cards after every game but used the remaining cards to deal for a new game until none were left. That way, if a player memo-

---

* These strict rules are intended to prevent employees from collaborating with the players in order to split the winnings after deliberately losing.

rized which cards had been used and changed his strategy accordingly he should be able to increase his probability of winning, if he knew how to use this available information. For this, he would have to use the conditional probabilities of a card being dealt, given the reduced deck of cards. But the rules should be simple and easily remembered to enable the player to decide quickly whether to ask for another card or not. Thorp, with the help of MIT's IBM 704 computer, worked out* an easily memorizable strategy which provided some advantage against the casino. At the meeting of the American Mathematical Society in Washington in 1960 he gave a talk on his computations.

In a couple of days, he received a letter from a businessman who offered him 100 thousand dollars to try his system out. Thorp accepted the offer and — after the businessman hadle arned his system — the two went to Nevada. The experiment was a complete success: the businessman won 17,000 dollars in two hours. The owner of the casino was not at all as enthusiastic as Thorp and his companion to see the conquest of science, and the next day, giving several excuses, he refused to let them play. Later, Thorp tried other casinos but his fame preceded him and they would not even let him in. At some casinos, he disguised himself with a false beard, or as a Chinese to get the tables but no disguise could camouflage his continuous winnings, and so he was forced to stop applying his mathematical results in real life. His revenge on the casinos was the publication of a book with a detailed description of his system (Ref. [9]). So many people learned the winning strategy that the casinos were forced to change the rules of the game. One of the changes introduced was that the entire deck of cards was shuffled after every game, thus removing the basis of the strategy. In this way, we have come back to our starting point: the importance of shuffling.

BIBLIOGRAPHY

[1] P. Erdős–P. Turán: On some problems of statistical group theory. I., *Zeitschrift für Wahrscheinlichkeitstheorie and Verwandte Gebiete.* **4,** 1965. 175–186.
[2] A. Prékopa–A. Rényi–K. Urbanik: O pregelnom raspregelenii dlya summ nezavisimih sluchaynih velichin na bikompaktnih kommutativnih topologicheskih gruppah, *Acta Math. Ac. Sci. Hung.* 1956. 11–16.
[3] U. Grenander: *Probabilities on Algebraic Structures,* Wiley, New York, 1963.
[4] K. Jordan: *Fejezetek a klasszikus valószínűségszámításból,* (On the Classical Probability Theory), Akadémiai Kiadó, 1955.

---

* It took 3 hours of computer time.

[5] E. Culbertson: *New and Complete Summary of Contract Bridge*. John C. Winston, Philadelphia, 1935.

[6] É. Borel–A. Chéron: *Théorie mathématique des bridges, à la portée de tous*. Gauthier-Villars, Paris, 1955. 1—424.

[7] J. L. Doob: *Stochastic Process*. Wiley, New York, 1953.

[8] L. E. Dubbins–L. J. Savage: *How to Gamble if You Must*. McGraw-Hill, New York, 1965.

[9] E. O. Thorp: *Beat the Dealer. A Winning Strategy for the Game of Twenty One*. Blaisdell, New York, 1962.

[10] G. de Rham: Sur quelques courbes definies par des équations fonctionelles. *Rendienti del Seminario Matem. Univ. Torino* **16.** 1956–57, 101–113.

Notes on the teaching of probability theory

In what follows, I will discuss the teaching of probability theory in general — regardless of the type of school or the age of the students.

Three basic questions will be considered:
1. why should probability theory be taught?
2. what should be taught? and
3. how should it be taught?

In other words I will speak of the aim, the content and the methods of teaching probability theory.

My remarks are based on personal experiences, not only from teaching at the University, but also from conducting courses on probability at the Free University of Budapest for interested high school students and from the television series I made which was watched by a very eclectic audience, from the very young to grown-ups, with quite diverse interests and background knowledge.

1. WHY SHOULD PROBABILITY THEORY BE TAUGHT?

First, it may seem that this question can be answered adequately only if we are considering a certain type and level of education. In my opinion, there are still some general points which can be made anyway. I will state what I think the main goals of teaching probability theory are. These goals should be kept in mind no matter which branch of probability theory one is teaching, although they might be stressed differently depending on the kind of school involved. They are as follows:

A) Probability theory should be taught because of the important role it can play in the development of the students' ability to think.

B) It should be taught because of its usefulness in everyday life, in science and technology, etc.

C) It is important, indeed indispensable in mathematical education.

I will now expand upon these points.

A) At the University in Budapest, there was a Professor of Law who used to ask the following question during examinations: "what do you see when you look down from Gellért Hill to the city?" The student was supposed to answer: "Subjects of law and legal entities"!

I do not know what this professor would have said had somebody answered "stochastic processes". He probably would not have understood the answer, although it would have been as correct as the one he was expecting. To understand the concept of probability is indispensable in comprehending the world around us; it is a corner-stone of our scientific world-view. Any branch of mathematics will help to further the mental development of students and teach them to think logically in clearly defined ideas. But the role of a student's thinking is more than that. Probability theory can teach students the usefulness of clear and logical thinking even if they have to deal with uncertainty (and uncertainty is what we encounter almost always in reality).

The study of probability even strengthens students' character. For example, it increases their courage when they understand that certain failures are due to chance so that a set-back is not a sufficient reason for giving up. Primitive people have a tendency to be very superstitious: if something goes wrong they try to attribute it to somebody's maliciousness even if such is not the case. The reason for this is that they do not understand the notion of chance. The study of probability theory can help to erase these remnants of magical thinking from the Stone Age, to make people more understanding toward fellow human beings and to help them to find their place in society.

B) In everyday life we encounter chance continuously. Probability theory can teach us how to take into account the different risks involved in different decisions and then arrive at a reasonable attitude. Choosing the most suitable insurance policy from among those available is a good example of a situation in which probability theory can be applied in our lives. In preparing the family budget or planning a trip, we have to estimate costs which, to a certain extent, depend on chance. These examples show that everybody needs to know the laws of chance.

The application of probability theory in science, technology, economics, etc. has risen so much in importance that more and more people need a working knowledge of it in their work. How much weight we should give to this consideration depends on the type of school involved. We should, however, keep in mind that nowadays every educated person, independently of his/her profession, should have some knowledge of subjects such as atomic energy, radioactivity, genetics, etc., and to understand these at even a layman's level one needs a certain knowledge of probability theory. Today, when we learn the probability of rain for tomorrow during the weather forecast, it would be practical for everyone to know what that probability really means.

C) Familiarity with the element of probability fosters an understanding of the connection between reality and mathematical models. Those pupils who have not had any exposure to the question of probability during their education will never have an adequate conception of what mathematics is and what it can be used for. People who are not acquainted with probability theory share a common misunderstanding, namely, that mathematical methods can be applied only to situations where a simple and accurate dependence exists among a few precisely measurable entities. One often hears, even now, that mathematical methods cannot be used to describe certain events because of their complexity. This is a prejudice of people who have learned some mathematics but not probability theory and this point of view has (at least in some countries) held up the application of mathematical methods in economics, sociology, biology, psychology and other areas for quite some time.

It is worth noting that the idea of teaching probability theory at high school or elementary school accords with other modern trends in teaching mathematics. On the one hand, some knowledge of set theory and Boolean algebra makes it easier to teach probability theory; on the other hand the study of probability theory will, by instructive applications, enhance the understanding of the notions of set theory and Boolean algebra.

2. WHAT SHOULD BE TAUGHT?

I will make just a few remarks on this topic because my aim is to deal in the main with general issues which are of importance at any educational level, while a specific course outline would depend very much on the kind of school involved, the age of the pupils, their mathematical background, etc.

I think that every course on this subject should contain some material on each of the following themes:

A) Probabilities in reality, i.e., the demonstration of statistical laws in everyday life, in nature, in games of chance.

B) The mathematical theory of probability.

C) The application of probability theory in the description of random mass phenomena and in the prediction of possible outcomes.

D) The history of probability theory, including a discussion of the philosophical questions relating to the notion of probability. The order of these four points corresponds, in my view, to the logical order in which they should be discussed.

I want to avoid any misunderstanding that may arise when I say that the teaching of probability theory should begin with the introduction of the students to the idea of statistical laws. I do not mean that one should start with statistics. On the contrary, I have not found any of the attempts to teach statistics without the critical notions of probability as prerequisites satisfactory either from a logical or a pedagogical point of view. It would be desirable to start the teaching of probability theory by discussing well chosen examples and experiments. First, we should indicate what are the basic issues to whose understanding the mathematical theory of probability can contribute and only then should probability theory be tailored to the particular age group and mathematical background of the students. Of course, the length of the school term and the particular goals of the school (if any) should also be taken into account. In my opinion, mathematical statistics should be taught as a separate subject only at the university level to those students who need it.

I have encountered opinions to the effect that the practical importance of probability theory can only be demonstrated by way of teaching statistics. I myself do not think so; a great part of the most important applications can be understood with some introductory knowledge of probability theory alone. It should be pointed out in any course on probability theory that, in practice, the basic parameters for most of the cases ought to be determined empirically, and that, if one has a large enough sample, such a task will not require difficult statistical procedures. In the introductory discussion, it ought to be explained that the study of the inverse problems of probability theory (where we want to infer the parameters of a probability distribution from observations) is the subject of another field; mathematical statistics, which is based on probability theory, yet is not a part of it but an independent subject. The Bayes method can be discussed in the context of probability theory, so that, if time permits, part of it can be included in the introductory lectures.

Concerning point D), I think that while it is generally desirable and useful to discuss the historical background of any subject, such a discussion is especially useful in the teaching of probability theory. Even in a short introductory course, it is important to mention the philosophical problems relating to the concept of probability because it will help the students learn the particular method of thinking used in probability theory. These philosophical questions can be discussed during the historical overview: this will constitute the history of the philosophy of probability.

Lastly, I would like to emphasize that I consider *entropy* and *information* to be fundamental notions in probability theory and I would like to suggest that teachers of probability theory include some discussions of these notions in their courses.

3. HOW SHOULD PROBABILITY THEORY BE TAUGHT?

The difficulties relating to this question are exactly the opposite of those discussed in relation to point 2. Here one can say so much at any particular teaching level without restricting oneself that some selection is necessary. Putting aside many important questions, I wish to make some remarks only about the following three:
  A) The question of mathematical precision.
  B) Experiments concerning random events.
  C) The introduction of the concept of probability space.

A) In general, I advocate reasonable precision in the teaching of mathematics because I feel that without precision, mathematics is not mathematics. This does not mean that every statement should be proved: some theorems can be given without any proof; others can be substantiated by heuristic reasoning and only a few need to be proved. Still, we should make each situation very clear so that the students will always know what has been proven and what has not. Particular care should be taken not to call a proof what is only heuristic reasoning. Similarly, we should clearly distinguish between definitions and theorems. These remarks apply to the teaching of any branch of mathematics, but I dwell on them because these basic rules are often violated in the teaching of probability theory. If the teacher wants the students to understand why precision is necessary, he/she can demonstrate by means of well chosen examples where imprecision leads to incorrect results. In general, well chosen examples ought to be the base of mathematical instruction. And no other branch of mathematics has such a great selection of exciting and still elementary examples as probability theory.

B) Statistical laws can be illustrated by data from books, newspapers etc., but students will be even more impressed if experiments carried out in front of (and preferably by) them, supply the necessary data. Some teachers do not agree with this approach because they fear that the experiments will not produce exactly the results they were expecting (and this can in fact happen because of the nature of these experiments). Such a fear, however, is unfounded and if the teacher understands probability theory thoroughly he/she cannot get into an uncomfortable position. Of course, the teacher has to react fast: evaluating unexpected results is always more difficult than explaining problems solved beforehand. The benefit of an experiment conducted in the school with the students is so great that I support it regardless of such difficulties. These experiments should be planned carefully. For example, in my talk on TV, I wanted to do Buffon's famous needle experiment. I was astonished that, although most books on probability theory mention it, none of them provides any practical advice as to how to conduct it so that the fundamental conditions will be satisfied. In the end, I had to construct a simple mechanism myself for this purpose. I had a similar experience with another classical experiment, the Galton-desk. I have learned that if one does not conduct this experiment carefully enough, the results will differ significantly from what one expects because the deflections of the balls into the different rows are strongly interdependent. In this case, I had to construct a special device in order to end up with the desired result. As far as dice are concerned, I have found the icosahedron (made in Japan for quality control purposes) to be the best. As far as I know, these are produced in great quantities in Japan. I do not think it would be too difficult to produce reliable normal dice for the purpose of teaching probability theory. During my television talks, I conducted experiments not only with simple dice but also with the kind of bones used by the ancient Greeks and Romans. There are, of course, many other simple experiments that are suitable for conducting in school: tossing a coin, selecting cards from well shuffled decks, roulette, etc. I realized what a great help examples derived from games of chance can be when my colleague Dr. Pál Révész told me about his difficulties in teaching probability theory in Ethiopia where games of chance are forbidden (so that he was advised not even to mention them). There are also certain methods employed in quality control that can be used for demonstrations in the classroom. For example, in one of may talks on TV, I carried out an experiment on a sack of small plastic balls using a plastic shovel indented with 100 holes in a square 10×10 arrangement. When I put the shovel among the balls, they were attracted to it electrostatically and filled the holes. The balls were of two colors; most were white and a small fraction (1/4) were red. Thus the results approximated the Poisson distribution quite well.

    I would like to point out that the data obtained in an experiment can be analysed in many ways and a particular analysis can lead to much more than an understanding of the idea of statistical relations. Specifically, it can lead to the concept of independence and other, less obvious concepts related to random events. For example, on many occasions I have shown students two sequences of zeros and ones, telling them that one was the result of tossing a coin (with 0 representing heads and 1 tails), and that the other was only an artificial random sequence. Both sequences had about 150 elements and the students had to guess which one was which. (In general, the artificial sequence was quite regular, having no long series of constant ones or constant zeros while in the actual random sequence, they did, of course, occur.)

C) Finally, I would like to relate how I recently introduced the concept of probability space. At first glance, it would seem that this is only a terminological innovation, but I will show that there is more to it.

What is usually called probability space*, namely a triple $(\Omega, \mathcal{A}, P)$ of a non-empty set Ω, a σ-algebra $\mathcal{A}$ of sets contained in Ω, and a probability measure P on $\mathcal{A}$ for which $P(\Omega)=1$, I call an *experiment*, and the elements $\omega \in \Omega$ are called the possible *outcomes* of an experiment. Any subset A of Ω is called an *event*. An event consists of the possible outcomes contained in it as a set. Those subsets of Ω which are elements of $\mathcal{A}$ are *observable events*, while those that do not belong to $\mathcal{A}$ are considered unobservable. As usual, I interpret $P(A)$ as the probability of event A. It must be stressed that we do not in any way define the probability of unobservable events.

A typical example is tossing two identical dice. In this case, Ω consists of 36 pairs of numbers, $(a, b) \in \Omega$ where $1 \leq a \leq 6$ and $1 \leq b \leq 6$. $\mathcal{A}$ consists of subsets A of Ω such that if $(a, b) \in A$ then $(a, b) \in A$. There are 2^{21} such subsets out of the total 2^{36}; these are the observable ones.

This example shows that even if Ω is finite, it is not always suitable to include in $\mathcal{A}$ all of the subsets of Ω. Of course, if instead of the set of ordered pairs (a, b), we take the set of inordered pairs to be Ω, then $\mathcal{A}$ will be the set of all subsets of Ω. In general, it is more practical to choose the relatively large set of possible outcomes and limit the number of sets having defined probabilities.

Thus we arrive logically at the assumption that $\mathcal{A}$ has to be an algebra of sets, since it is obvious that, if an event is observable then its opposite is also observable and, moreover, if two events are observable then the event consisting of the outcomes belonging to at least one of these events is observable too. (This is not true in quantum mechanics but is true for "classical" observations.)

I have tried to introduce the notion of probability space this way in different schools at the university and high school level and my experience is that students understand this concept more easily if the accent is on observability. The advantages of this method become obvious later: it makes it easier to understand general concepts like conditional probability and conditional expected value if from the beginning students get used to the fact that the system of sets with which a probability is associated is necessarily the largest possible one. Many textbooks explain the fact that the probability measure is defined on a σ-algebra of subsets in the sample space (instead of all sub-

---

* The following two pages are primarily directed to specialists. The required definitions can be found for example in A. Rényi: *Probability Theory* (Akadémiai Kiadó, Budapest, 1970). (Gy. Katona)

sets) by stating that often it is impossible, by pure mathematical reasoning, to extend the measure to all subsets. Although this last fact is, of course, true, I think that such an explanation is misleading. Usually an extension of this sort would be completely meaningless because the extended measure would lose its original probability theoretical meaning. I do not want to go into detail on this question since it comes up only in higher level courses in probability theory. To go back to the more elementary level, I would like once again to emphasize my own experience: the concept of probability (or, more precisely, the mathematical structure of probability theory) can be more easily understood if, from the beginning, we introduce the idea of the observability of those events for which probability is defined.

Variations on a theme by Fibonacci

In music, composers quite commonly compose variants, "variations" on a theme. Mozart liked this form very much; the first movement of his Sonata in A-major (K. 331) consists of a set of variations, as does the first movement of Beethoven's Sonata in A-flat minor (op. 26). The main characteristic of this form is that the composer starts with a simple, basic theme and creates variations on it which differ, sometimes quite strikingly, in rhythm and mood from their original. No matter how surprising these variations are, the listener feels that each particular one existed as a possibility in the theme: it only required an ear (i.e., that of a composer) sensitive enough to "hear" it and give it life.

In what follows, we will try to follow the example of music-literature by introducing a simple mathematical theme — the so-called Fibonacci sequence — followed by its numerous "variations". These variations are based on the different characteristics, interpretations, applications and generalizations of the Fibonacci sequence.

THE THEME

Let us consider the following sequence:

$$1, 2, 3, 5, 8, 13, 21, \ldots .$$

What kind of pattern can we discover in it? It is easily noticed that beginning with the third term, every term of the sequence is equal to the sum of the two preceding, i.e., $3=1+2, 5=2+3, 8=3+5, 13=5+8, 21=8+13, \ldots$. Seeing this, we can continue the sequence:

(1) $\qquad 1, 2, 3, 5, 8, 13, 21, 34, 55, 89, 144, 233, 377, 610, \ldots .$

Variation 1. Try to continue the sequence (1) backwards, preserving the property that any term is equal to the sum of the two preceding. The sequence can be indefinitely extended to the left as follows: to the left of what is now the first element of the sequence, write the number obtained by subtracting it from the second element of the sequence. Proceeding in this way, we end up with a sequence which is infinite in both directions:

(2) $\quad$ ..., −21, 13, −8, 5, −3, 2, −1, 1, 0, 1, 1, 2, 3, 5, ...

Notice that the numbers to the left of zero are the same as those to the right, but with alternating sign.

Variation 2. The rule underlying the construction of (1) can be expressed in a different way. Starting with the second term of the sequence, write under each number the difference between it and its predecessor. In this way, we obtain the same numbers in the lower row as in the upper row, but shifted one position to the right:

$$1, 2, 3, 5, 8, 13, 21, 34, ...$$
$$1\ 1\ 2\ 3\ \ \ 5\ \ \ 8\ 13\ ... .$$

This is also true of (2):

$$..., -8, 5, -3, 2, -1, 0, 1, 1, 2, 3, 5, 8, ...$$
$$13, -8, 5, -3, 2, -1, 0, 1, 1, 2, 3, 5, ...$$

Variation 3. Sequence (1) is called the Fibonacci sequence in the literature, because it first appeared in a mathematical work entitled *Liber Abaci* (1202) by the Italian mathematician Leonardo Fibonacci (=son of Bonacci) (1170–1250). Maestro Leonardo (who was called "the Pisan" because of his birthplace) had travelled in the East as a merchant and based his book on the work of Arab mathematicians such as Al-Khawarizmi, Abu Kamil and others. It is well known that the works of the ancient Greek mathematicians were forgotten in the Middle Ages in Europe, and only survived in the works of the Arab mathematicians who were also inspired by the mathematicians of India. At the time of the Crusades, the forgotten mathematics of the ancients was brought to light again in Europe through the works of the Arab mathematicians. The first important European mathematical textbook was the above mentioned book by Fibonacci. It is hard now to tell how much of it he had learned from the Arabs and how much was his own. In his book, there are a number of examples (such as sequence (1)) whose sources are unknown. We do not know whether Maestro Fibonacci created these or whether he took them from other source-books which have not survived. Fibonacci presented sequence (1) as the solution of a problem concerning the *reproduction of rabbits*.

We will rephrase the problem in terms of the *growth of* trees (the assumptions being more realistic in that situation than with rabbits). Assume that a tree grows in the following way:

Each new branch just grows during its first year, but gives birth to a new branch every year starting from the second year of its existence. The question is: how many branches will a tree planted today as one branch have in 1, 2, 3, 4, ... years. The assumed pattern of growth of the tree is shown in Fig. 1.

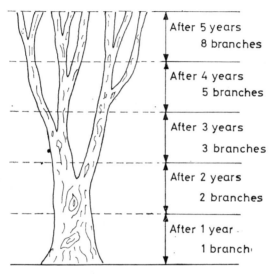

Fig. 1

The tree will have 1 branch in the second year, 2 branches in the third, 3 in the fourth, 5 in the fifth and 8 in the sixth: writing these numbers in a sequence, we obtain (1). Now we can easily see why the rule that every term of the sequence equals the sum of the preceding two holds. We can calculate the number of branches in any given year if we add the number of branches in the previous year to the number of newly grown branches. This latter number is equal to the number of branches which are at least 2 years old since only from these can a new branch grow.

Variation 4. In a new housing development, the apartment houses are to be painted so that each level (including the ground floor) will be either blue or white. For aesthetic reasons, no two consecutive levels are to be blue. In how many ways can a particular building be painted, given that it has a fixed number of stories? Fig. 2 shows all the possibilities for one, two, three and four-storey houses.

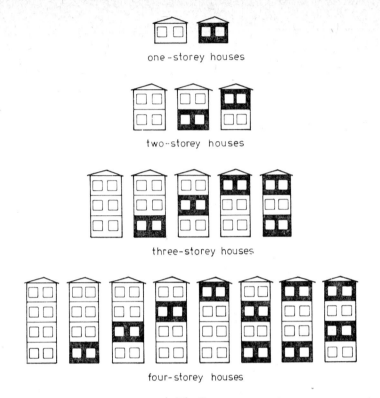

one-storey houses

two-storey houses

three-storey houses

four-storey houses

Fig. 2

There are two, three, five, and eight patterns in case of one, two, three and four-storey houses, respectively. As the number of levels of the building increases, the numbers of possibilities are as follows:

(3) $\qquad 2, 3, 5, 8, \ldots$

which corresponds to the Fibonacci sequence, without its first term. Is the sequence true for taller buildings, for example are there 13 possibilities for a 5-storey building? The answer is yes and this can be demonstrated as follows. In a 5-storey building, the fifth level can be either blue or white. Keeping in mind the constraint that "no two consecutive levels can be blue", there are as many patters in which the 5th level can be painted white as there are ways of painting a 4-storey building (namely 8), and there are as many patterns in which the 5th level can be blue as there are patterns in which the 4th level of a 4-storey building is white, which is exactly the number of patterns for which the painting of a 3-storey building (namely 5). Hence, there are $8+5=13$ ways of painting a 5 level house to comply with the restriction of "no two blues together". Similarly, one can see that the number of patterns corresponding to six-storey building is $8+13=21$, to seven-storey building $13+21=34$, etc. The solution of the problem is the Fibonacci-sequence.

Variation 5. A TV-station has the policy of broadcasting a particular program (for example, a mathematical talk series for laymen) on certain given days of the week and not on others. How many different weekly schedules can be made if this program is not to be shown on two consecutive days? Notice that although this variation is quite similar to 4, it is not identical to it. If we assign the seven levels of a seven-storey building to the seven days of the week (so that Sunday corresponds to the ground floor, Monday to the second storey, etc.) and the color blue to the above-mentioned program, then it is true that every weekly schedule constructed to satisfy the given condition will correspond to a possible color pattern of the 7-storey building, but the converse is not true. If, for example, both the ground and the seventh floors are blue, that would correspond to having the program screened on Sunday and Saturday, which is not allowed. The possible schedules are shown in Fig. 3, where the days of the week are represented by the numbers 1 to 7 written around the edge of a circle, and the days on which the program can be shown are circled. Counting all of these we end up with a total of 29. This number is not in the Fibonacci-sequence; still, this problem has a connection to the Fibonacci-numbers. Let us see what happens if we want to air the program on a 3 or 4-day schedule (instead of the 7-day basis). Let us reformulate the problem illustrated by Fig. 3 as follows. The guests at a banquet are to be seated at round tables. All tables accommodate the same number of people and the seats are numbered. How many ways are there of assigning seats to guests if no two women can sit next to one another? In Fig. 4 we can see the possibilities for tables seating 2, 3, 4, 5 and 6 guests. The women guests sit in the places marked by circled numbers. According to Fig. 3, there are 29 possible seating patterns in the case of a table for 7. For tables seating 2, 3, 4, 5, 6, and 7, the numbers are:

(4) $\qquad$ 3, 4, 7, 11, 18, 29,

respectively. In order to establish the connection between (4) and the Fibonacci sequence, let us investigate why 11 is the number of possible ways of seating five around a table.

Seat number 1 can be occupied by a man or a woman. If the person is a man the number of possible seating arrangements for the remaining four places is the same as the number of ways a 4-storey building can be painted white and blue, i.e., 8. In the case where a woman is seated in the first seat,

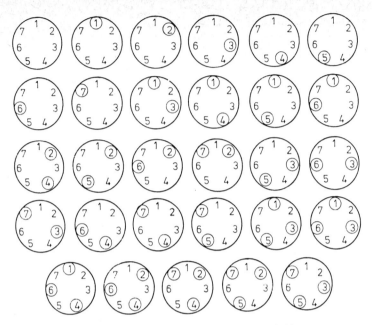

Seating patterns for seven around a table

Fig. 3

she will have to be seated between two men; the remaining two places will give as many different possible arrangements as there are patterns for the painting of a 2-storey building, i.e., 3. The total number, then, is $8+3=11$. Similarly, one can see that the number of possible seating arrangements at a table for 6 is $13+5=18$, that at a table for 7 it is $21+8=29$, etc.

Sequence (4) is therefore derived from the Fibonacci sequence by adding to every term (from the third term on) the one before the preceding $2+1=3$, $3+1=4$, $5+2=7$, $8+3=11$, $13+5=18$, $21+8=29$, etc. If we have a table for 10, there are 123 ways of seating guests.

We should note that the case of a table for one can be included in the sequence only under the condition that no woman can sit alone; then the general rule applies to give $1+0=1$ possible seating pattern. The rule for the pattern of seatings can be explained to have the meaning that the men are to entertain the women during supper and that that is why we prescribe that a man must sit on the right and left of each woman. Whoever sits at a table for one will not have any neighbour (or we could say that he his own neighbour). Accordingly, a woman cannot sit by herself.

Although this problem has led us to the Fibonacci numbers, its solution is

Seating patterns for two around a table

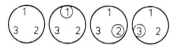
Seating patterns for three around a table

Seating patterns for four around a table

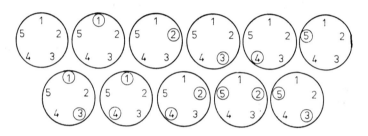
Seating patterns for five around a table

Seating patterns for six around a table

Fig. 4

not simply constituted by the terms of a Fibonacci sequence but rather by the sums of two of its terms two places apart.

Variation 6. Let us set out again the sequence we arrived at in Variation 5.

(5) $\qquad$ 3, 4, 7, 11, 18, 29, 47, 76, 123, ...

Looking at the sequence of differences between each two consecutive terms, we realize that the resulting sequence has the same property as the Fibonacci sequence, namely that every term is equal to the sum of the two preceding ones:

$$7 = 3+4, \quad 11 = 4+7, \quad 18 = 7+11, \quad 29 = 11+18,$$
$$47 = 18+29, \quad 76 = 29+47.$$

Why? What is the connection between this feature of (5) and its origin in sequence (1)? To see this, it will be useful to adopt the usual mathematical notation for sequences. Denote by F_n the n-th term of sequence (1) so that

$$F_1 = 1, \quad F_2 = 2, \quad F_3 = 3, \quad F_4 = 5, \quad F_5 = 8,$$

and so on.

According to the definition of the Fibonacci numbers we have found that for every n

(6) $\qquad F_n = F_{n-1} + F_{n-2},$

holds. Denote by G_n the number of possible seating arrangements around a table for n, i.e.,

$$G_1 = 1, \quad G_2 = 3, \quad G_3 = 4, \quad G_4 = 7, \quad G_5 = 11, \quad \text{etc.}$$

Previously we saw that

(7a) $\qquad G_n = F_n + F_{n-2}.$

We want to show that

(7b) $\qquad G_n = G_{n-1} + G_{n-2}.$

From (6) and (7a), it follows that

$$G_n = F_n + F_{n-2} = (F_{n-1} + F_{n-2}) + (F_{n-3} + F_{n-4}) =$$
$$= (F_{n-1} + F_{n-3}) + (F_{n-2} + F_{n-4}) =$$
$$= G_{n-1} + G_{n-2}.$$

Variation 7. We have already arrived at two different sequences where every term is equal to the sum of the two preceding it. Let us now try to review all such sequences. Let us call a *Fibonacci-type sequence* any sequence in which any term from the third term on is the sum of the two preceding it. If we denote by a_n the n-th term of the sequence $(n=1, 2, ...)$, then the above condition can be written as follows

(8) $$a_n = a_{n-1} + a_{n-2} \quad (n = 3, 4, ...).$$

Obviously, the first two terms of such a sequence can be chosen arbitrarily since (8) does not constrain either the first or second terms. On the other hand, once the first two terms are decided on, the rest of the sequence is determined and the whole series can be calculated. For example, by choosing 1 and 6 to begin the sequence we arrive at the sequence

(9) $$1, 6, 7, 13, 20, 33, 53, 86, ...$$

Of course, a Fibonacci-type sequence is also determined if any two terms other than the first two are given.

For example, let's find the Fibonacci-type sequence having 1 as its first term and 9 as its fourth. Let x denote the second term. The third term is $1+x$, the fourth $x+(1+x)=1+2x$. If our condition that the fourth term is 9 is to be met, the equation $1+2x=9$ must be satisfied; hence $x=4$. The sequence we were looking for is:

(10) $$1, 4, 5, 9, 14, 23, ...$$

Variation 8. Let us see which of the simple sequences one learned about in school can be classified as Fibonacci-type sequences. It can quite easily be seen that an arithmetic sequence can be of the Fibonacci-type only if all its terms are equal to 0 since an arithmetic sequence has the property that the difference between any two consecutive terms is constant, while the differences between any two terms of the Fibonacci-type sequence (as we have seen in Variation 2) result in the same sequence (shifted to the right by one position).

Let us now examine if a geometric sequence can be considered to be a Fibonacci-type sequence. The geometric sequence has the property that the quotient of any two consecutive terms is constant.

If $a_1, a_2, ..., a_n, ...$ is a geometric sequence, and q is the quotient, then $a_2 = a_1 q$, $a_3 = a_2 q = a_1 q^2$, and so on; in general,

(11) $$a_n = a_1 q^{n-1} \quad (n = 1, 2, ...).$$

If a geometric sequence (a_n) is of the Fibonacci-type as well, then $a_3 = a_1 + a_2$, $a_4 = a_2 + a_3 ... , a_n = a_{n-1} + a_{n-2}$ should also hold. In other words, according to equation (11) the following relationships:

(12)
$$a_1 q^2 = a_1 + a_1 q$$
$$a_1 q^3 = a_1 q + a_1 q^2$$
$$\dots\dots\dots\dots\dots\dots\dots\dots\dots$$
$$a_1 q^{n-1} = a_1 q^{n-2} + a_1 q^{n-3}$$

must be satisfied. Any one of these equations beginning with the second can be derived from the preceding one by multiplying the right and left by q. If, therefore, the first equation in (12) is true, the rest will be true as well: our result, then, is that a geometric sequence will be of the Fibonacci-type if and only if

(13) $$a_1 q^2 = a_1 + a_1 q.$$

Assuming that a_1 is not zero, we arrive at

(14) $$q^2 = 1 + q.$$

The necessary and sufficient condition for the geometric sequence (11) to be of Fibonacci-type is that q satisfy (14). Equation (14) has two roots:

(15) $$q_1 = \frac{\sqrt{5}+1}{2} \quad \text{and} \quad q_2 = -\left(\frac{\sqrt{5}-1}{2}\right).$$

If, therefore, q_1 and q_2 are numbers as defined in (15), then the geometric sequences $a_n = a_1 q_1^{n-1}$ and $a_n = a_1 q_2^{n-2}$ are also Fibonacci-type sequences where a_1 is an arbitrary number.

Variation 9. Since $\sqrt{5}$ is irrational, q_1 and q_2 will also be and one will obtain Fibonacci-type sequences with non-integer terms. If, for example, $a_1 = \sqrt{5} + 3$, then the Fibonacci-type sequence $a_n = a_1 q_1^{n-1}$ ($n = 1, 2, \dots$) is as follows:

(16) $$\sqrt{5}+3, \quad 2\sqrt{5}+4, \quad 3\sqrt{5}+7, \quad 5\sqrt{5}+11, \dots.$$

Note that each term of this sequence is calculated by multiplying the appropriate term of the Fibonacci-type sequence 1, 2, 3 by $\sqrt{5}$ and by adding to the result the appropriate term of the Fibonacci-type sequence 3, 4, 7, 11, ...

It can easily be seen that multiplying the terms of a Fibonacci-type sequence by a constant will result in a Fibonacci-type sequence, and that adding two arbitrary Fibonacci-type sequences also produces such a sequence. In general, if a_n and b_n are two arbitrary Fibonacci-type sequences, and A and B are arbitrary numbers, then the sequence

(17) $$C_n = A a_n + B b_n \quad (n = 1, 2, \dots),$$

will also be of the Fibonacci-type. That is,

$$\text{if } a_n = a_{n-1}+a_{n-2}, \text{ and } b_n = b_{n-1}+b_{n-2},$$

then from (17) we can conclude that:

(18) $$C_n = Aa_n+Bb_n = A(a_{n-1}+a_{n-2})+B(b_{n-1}+b_{n-2}) =$$
$$= Aa_{n-1}+Bb_{n-1}+Aa_{n-2}+Bb_{n-2} =$$
$$= C_{n-1}+C_{n-2}.$$

This means that, with the help of two appropriately chosen Fibonacci-type sequences, any Fibonacci-type sequence can be generated. Thus the following statement is true: any Fibonacci-type sequence (a_n) can be written in the form:

(19) $$a_n = AF_n+BG_n,$$

where (F_n) is the sequence $(1, 2, 3, 5, 8)$ and (G_n) is $(1, 3, 4, 7, 11, ...)$.

If a_1 and a_2 are given, we can choose constants A and B such that (19) will hold for $n=1$ and $n=2$. To the end, A and B should be chosen to satisfy the following two equations:

(20) $$a_1 = A+B,$$
$$a_2 = 2A+3B.$$

The solutions are:

(21) $$A = 3a_1-a_2,$$
$$B = a_2-2a_1.$$

Our result, then, is as follows: any arbitrary Fibonacci-type sequence can be written in the form

(22) $$a_n = (3a_1-a_2)F_n+(a_2-2a_1)G_n.$$

Formula (22) also shows how, with the help of the base-sequences F_n and G_n, an arbitrary term of a Fibonacci-type sequence can be expressed in terms of its first two terms. For example, according to (22), sequence (10) whose first two terms are $a_1=1$ and $a_2=4$, can be written as:

(23) $$a_n = 2G_n-F_n.$$

Thus, for example, the 5th term of the sequence $a_n=2G_n-F_n=2 \cdot 11-8 = 14$, which is indeed the fifth term of the above mentioned sequence $1, 4, 5, 9, 14, 23, ...$.

Similarly, any Fibonacci-type sequence a_n can be expressed with the help of the Fibonacci-type sequences q_1^{n+1} and q_2^{n+1}, where $q_1 = \dfrac{\sqrt{5}+1}{2}$ and $q_2 = -\left(\dfrac{\sqrt{5}-1}{2}\right)$. For the Fibonacci-sequence F_n itself we get

$$F_n = \frac{q_1^{n+1} - q_2^{n+1}}{\sqrt{5}},$$

i.e.,

$$F_n = \frac{(1+\sqrt{5})^{n+1} - (1-\sqrt{5})^{n+1}}{2^{n+1}\sqrt{5}}.$$

This formula is called the *Binet formula*.

Variation 10. Let us look at the quotients of consecutive terms in the Fibonacci-sequence (1): $\dfrac{2}{1}=2$, $\dfrac{3}{2}=1.5$, $\dfrac{5}{3}=1.666...$, $\dfrac{8}{5}=1.6$. If we continue this sequence, the quotients alternately decrease and increase but with smaller and smaller fluctuations.

It can be shown that the numbers $\dfrac{F_{n+1}}{F_n}$ approach the limit

(25) $$q_1 = \frac{\sqrt{5}+1}{2} = 1.6180... \,.$$

There are several ways to demonstrate this fact.

As we have shown, q_1 satisfies $q_1^2 = 1 + q_1$. Dividing this equation by q_1, we get

(26) $$q_1 = 1 + \frac{1}{q_1}.$$

On the other hand, from $F_{n+1} = F_n + F_{n-1}$ it follows that

(27) $$\frac{F_{n+1}}{F_n} = 1 + \frac{1}{\left(\dfrac{F_n}{F_{n-1}}\right)}$$

and therefore

(28) $$\frac{F_{n+1}}{F_n} - q_1 = -\frac{F_{n-1}}{F_n q_1}\left(\frac{F_n}{F_{n-1}} - q_1\right).$$

From (28), one can see that the quantities $\dfrac{F_{n+1}}{F_n} - q_1$ are alternately positive and negative numbers, and moreover, that if we denote by d_{n+1} the absolute value of $\dfrac{F_{n+1}}{F_n} - q_1$, then (considering that $F_{n-1} < F_n$),

(29) $$d_{n+1} < \dfrac{d_n}{q_1}.$$

This means that the deviation d_{n+1} is less than d_n/q_1. Since $q_1 = 1.6... > \dfrac{3}{2}$, every term of the sequence d_n is smaller than the previous one divided by $\dfrac{2}{3}$. Therefore d_n will be arbitrarily small if n is large enough, and so, the limit of $\dfrac{F_{n+1}}{F_n}$ will really be q_1 as n approaches infinity.

Variation 11. Where is the number $q_1 = \dfrac{\sqrt{5}+1}{2}$ familiar from? This is in fact a famous number. If an arbitrary straight line segment is divided into two segments so that the longer is the q-th part of the whole, then, after designating the length of the whole segment to be 1, the length of the parts will be $\dfrac{1}{q_1}$ for the longer, and $1 - \dfrac{1}{q_1} = \dfrac{q_1 - 1}{q_1}$ for the shorter.

Since $q_1 = 1 + \dfrac{1}{q_1}$, we can write

(30) $$\dfrac{1}{q_1} : 1 = \dfrac{q_1 - 1}{q_1} : \dfrac{1}{q_1}, \quad \text{i.e.,}$$

if a segment is divided into two in such a way that the greater part is $\dfrac{1}{q_1}$ of the whole, then the greater part is to the whole as the smaller part is to the greater. This kind of division was called *golden section* in ancient Greek mathematics.

The first application of the golden section first appeared in connection with the construction of the sides of a regular decagon. The side of a regular decagon, as can be seen from Fig. 5, is the q_1-th part of the radius of the circumscribed circle.

It can also be seen from this figure that the radius of the circle and the side of the decagon cannot be measured in terms of each other's length, i.e., the proportion resulting from the golden section is an irrational number. (This can also be seen from the formula $q_1 = \dfrac{\sqrt{5}+1}{2}$, since 5 is not a perfect square.)

We do not have space here to deal with the importance of golden section,

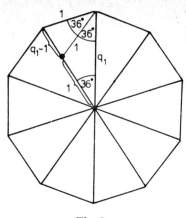

Fig. 5

but we should note that it played an important role in classical Greek art. One can see this proportion which, so it seems, is exceptionally pleasant to the eye, in many buildings and sculptures.

Variation 12. Let us investigate the following practical problem: at what speed does a car run most economically, i.e., at which speed is the car's gasoline consumption per 100 km the smallest possible. If this consumption (i.e., the amount of gasoline in litres needed to cover a distance of 100 km) is depicted as a function of speed (measured in km/hour) a graph such as that set out in Fig. 6 results.

This curve has one minimum (lowest point); it decreases until that point,

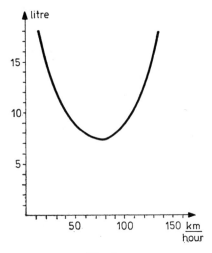

Fig. 6

and after that it increases. We want to find the value x, at which the rate of consumption is the smallest, i.e. we want to find the lowest point of the curve. Of course, we do not have the curve but we could measure the consumption for different speeds. This would be quite a lengthy procedure, especially if we want to measure accurately. Therefore it is desirable to find the most economical speed with a prescribed accuracy while making the smallest possible number of measurements. It can be shown that if we want to determine the most economical speed between bounds a and b by making n measurements with the smallest possible margin of error, then, as a first step, the given interval should be divided into F_n equal parts. The points where the first two measurements are taken are the F_{n-1}-st and the F_{n-2}-nd division points; after this, the process is repeated using the new interval defined by the two division points mentioned. (If the number of measurements is large enough, this process results in a golden section.) For example, let us assume that we want to find the optimal speed in 5 steps. Let us start from an interval which surely includes the most economical speed such as the interval 0 km/hour–160 km/hour. This interval should be divided into $F_5 = 8$ equal subintervals and we will take the first measurements at points $F_3 = 3$ and $F_4 = 5$, i.e., 60 km/hour and 100 km/hour. Let us assume that the measured consumptions at these speeds were $f(60)$ and $f(100)$, respectively. It can easily be seen that if $f(60) \geqq f(100)$, then the optimal speed is between 60 and 100 km/hour, while if $f(60) \leqq f(100)$, it should be between 0 and 100 km/hour. Assume that $f(60) < f(100)$. There can now only be 4 more measurements made and accordingly the interval 0–100 should now be divided into 5 equal parts. According to the method prescribed, measurements are to be taken at the second and third division points, since $F_2 = 2$, $F_3 = 3$.

The second division point is 40 km/hour. The third division point is 60 km/hour, but we already know the value of the function there. Assume that we have measured $f(40)$, and that $f(40) < f(60)$. We can accordingly be sure that the optimal speed is between 60–80 km/hour. We now need to carry out only one more measurement. Dividing the segment 60–80 km/hour into two (since $F_2 = 2$), we will measure the gasoline consumption $f(70)$ at 70 km/hour. If the result is $f(70) < f(60)$, then we will know that the optimal speed is between 70 and 80 km/hour. It can be proven that the best method for determining the desired optimal speed in five measurements is the one given above.

The preceding problem belongs to the field of mathematical search theory and as we have described, its solution leads to the Fibonacci sequence.* The

---

* This interesting application of the Fibonacci numbers was discovered by I. Kiefer. See *Proceedings of the American Mathematical Society* **4** (1953) 502.

practical meaning of the function whose maximum (or minimum) we want to determine is, of course, irrelevant. If, for example, we want to find the *optimal speed of rotation* for a power loom or the *most effective investment plan* we should go about doing so in the same way.

Variation 13. Let us start with the well-known Pascal triangle:

```
            1
           1 1
          1 2 1
         1 3 3 1
        1 4 6 4 1
       1 5 10 10 5 1
      1 6 15 20 15 6 1
     1 7 21 35 35 21 7 1
     . . . . . . . . . . . . . . . . . . . . .
```

Fig. 7

As is generally known, the k-th element ($k=0, 1, \ldots$) of the n-th row ($n=0, 1, 2, \ldots$) of the Pascal triangle gives the number of ways one can select k objects out of n, i.e., the binomial coefficient $\binom{n}{k}$. The distinguishing feature of the Pascal triangle is that any element in a row is the sum of the two consecutive elements immediately above it in the preceding row (e.g., $21 = 6 + 15$).

This rule is very similar to the rule generating the Fibonacci sequence. It can in fact be shown that there is a connection between the Pascal triangle and the Fibonacci numbers.

If we rewrite the rows of the Pascal triangle so that every number is written one row lower than its neighbour on the left in the original version, i.e., the elements of one original row are placed according to the knight's moves in chess in the new arrangement (two to the right, one down), then the row sums (the sums of the elements in each row) of the resulting "skewed" Pascal triangle are the Fibonacci numbers (see Fig. 8)

```
            1                row-sums: 1
           1                          1
          1   1                       2
         1   2                        3
        1   3   1                     5
       1   4   3                      8
      1   5   6   1                  13
     1   6  10   4                   21
    1   7  15  10   1                34
   1   8  21  20   5                 55
```

Fig. 8

For example:

$$\binom{8}{0}+\binom{7}{1}+\binom{6}{2}+\binom{5}{3}+\binom{4}{4} = 1+7+15+10+1 = 34.$$

In general we can write:

(31) $\quad F_n = \binom{n}{0}+\binom{n-1}{1}+\binom{n-2}{2}+\binom{n-3}{3}+\ldots = \sum_{0 \leq k \leq \frac{n}{2}} \binom{n-k}{k}.$

Relation (31) can be justified by the generating rule of the Pascal triangle or by the combinatorial meaning of the binomial coefficients and the combinatorial interpretation of the Fibonacci numbers (see Variation 4). Specifically, it can easily be demonstrated that, in the case of an n-storey building, the number of possibilities of having k levels painted blue is equal to $\binom{n-k}{k}$ and therefore (31) holds.

Variation 14. Let us return to sequence (2). Here every even number is preceded by two odd ones. Taking into account the rule for calculating the sequence and the fact that the regularity just mentioned is true for the beginning of the sequence, we see that the sequence can be continued to any length so that every two consecutive odd terms are followed by an even one which in turn is followed by two odd ones, etc. Let us investigate the divisibility by 3 of sequence (2). Write below every term the remainder after dividing by 3:

(32) $\quad$ −21 13 −8 5 −3 2 −1 1 0 1 1 2 3 5 8 13 21 34 55
 0 1 1 2 0 2 2 1 0 1 1 2 0 2 2 1 0 1 1.

Clearly, in the second row the 8-term sequence 0 1 1 2 0 2 2 1 is repeated over and over again, i.e., this sequence is periodic. If now we look for the sequence of remainders when dividing by 4, we will obtain a similar result:

(33) $\quad$ −21 13 −8 5 −3 2 −1 1 0 1 1 2 3 5 8 13 21 34 55 89
 3 1 0 1 1 2 3 1 0 1 1 2 3 1 0 1 1 2 3 1.

We can show that if Fibonacci numbers are divided by any number N, the remainders always create a periodic sequence. In other words, for every integer N, one can find a number d_n such that $P_{n+d_N} - F_n$ is divisible by N for all n. ($d_2 = 2$, $d_3 = 8$, $d_4 = 6$ as we have seen.)

The proof is quite simple. Any term of the remainder sequence is one of the numbers $0, 1, \ldots, N-1$. Any remainder can therefore have one of N possible values, so there can be N^2 possible pairs of consecutive terms. But the same rule applies to the remainder sequence as to the Fibonacci-numbers, meaning that any term is equal to the sum of the two preceding terms if that sum is smaller than N. If the sum is greater than N, then subtracting N from it will give the next term of the sequence. (This is expressed by saying: every term in the remainder-sequence is equal to the sum "modulo N" of the two terms previous to it.) Therefore the sequence of the remainders still has the property that the whole sequence is determined by any two given elements. For example, to get the sequence of remainders resulting from division by 7, we start with the first two terms: (1, 2) and calculate any term by adding its two predecessors, checking whether this sum is greater than or equal to 7 and subtracting 7 if this is the case. The sequence will be:

(34)   1, 2, 3, 5, 1, 6, 0, 6, 5, 4, 2, 6, 1, 0, 1, 1, 2, 3, 5, 6, 0, ...

If we now consider a part of a remainder-sequence (after division by N) consisting of N^2+2 members and look at consecutive pairs of terms, we will find N^2+1 such pairs, not all of which can be different since there can be N^2 possible pairs out of N numbers. If, on the other hand, the same pair appears in two places in the sequence, then the sequence will continue the same way from both places. We have now proved that the sequence of the remainders resulting from the division by N of every term in a Fibonacci-type sequence* is periodic with a period of at most N^2. Moreover, if we take into account the fact that the pair 00 cannot appear in the sequence of remainders (for that, it would be necessary for every term including the first, i.e., 1, to be divisible by N) then the maximum length of a period will be N^2-1, i.e., $d_N \leq N^2-1$. (Note that $d_2=3=2^2-1$ and $d_3=8=3^2-1$.) In the example of division by 7, we have determined the place where the pair 1, 2 is first repeated. We see that $d_7=16$. (Notice that d_7 is not equal to $7^2-1=48$, but is a divisor of it.)

Since 0 appears in sequence (2) and the sequence of the remainders after division by N is periodic, in the latter sequence zero appears infinitely many times. This then means that, given any N, there are infinitely many terms in sequence (1) which are divisible by N.

An interesting problem is that of the divisibility among the terms of the F_n-sequence. We give here the most important result without proof; the number F_n is divisible by the number F_m ($1 \leq m < n$) if and only if $n+1$ can be divided by $m+1$. It follows that F_n can be a prime number only if $n+1$ is a prime number or if $n=3$. This does not mean that if p is prime, then F_{p-1}

---

* Many other interesting facts concerning Fibonacci-numbers can be found in *Chisla Fibonacci* by N. N. Vorobjov (2. edition, Nauka, Moscow, 1964, 1–70.).

is definitely prime; moreover, the following is still an unsolved problem: are there infinitely many prime numbers in a Fibonacci sequence?

Let us note that the proof given can be applied to any sequence consisting of the numbers $0, 1, \ldots, N-1$, where every term is calculated from a fixed number of preceding terms in accordance with a specific rule and that all such sequences are necessarily periodic. This fact plays an important role in *the generation of the* so-called *pseudo-random numbers*. When there is a need for "random" numbers for the Monte-Carlo method in calculations by computer, these "pseudo-random" numbers are what the computer generates. The reasoning set out above shows that an algorithm which creates a new term of this "pseudo-random" sequence by accomplishing some operations on a fixed number of preceding terms will necessarily generate a periodic sequence. Therefore it should be kept in mind that it is necessary to have a large enough period in order to be able to use the sequence as "random".

Variation 15. Consider now the partial sums of the Fibonacci numbers:

$$S_1 = 1$$
$$S_2 = 1+2 = 3,$$
$$S_3 = 1+2+3 = 6,$$
$$S_4 = 1+2+3+5 = 11,$$
$$S_5 = 1+2+3+5+8 = 19,$$
$$S_6 = 1+2+3+5+8+13 = 32, \quad \text{etc.}$$

If we now take the resulting sequence

(35) $\qquad\qquad 1, 3, 6, 11, 19, 32, \ldots$

and add two to each of its terms, we will end up with a sequence of $3, 5, 8, 13, 21, 34, \ldots$ which is the Fibonacci sequence from its third term on. This can be represented by the following formula:

(36) $\qquad\qquad S_n - F_{n+2} = 2.$

The sequence S_n is not of Fibonacci-type; rather, its generating rule is:

(37) $\qquad\qquad S_n = S_{n-1} + S_{n-2} + 2,$

that is, every term exceeds the sum of the two preceding terms by two. What if we leave out every second term of the Fibonacci sequence:

(38) $\qquad\qquad 1, 3, 8, 21, 55, 144, \ldots\ .$

What is the generating rule in this case? It can be seen that, from the third term on, every term is equal to the difference between the immediately preceding term multiplied by three and the second preceding term (i.e., $8 = 3 \cdot 3 - 1$, $21 = 3 \cdot 8 - 3$, $55 = 3 \cdot 21 - 8$, $144 = 3 \cdot 55 - 21$, etc.). If H_n denotes the n-th term of sequence (38), then

(39) $$H_n = 3H_{n-1} - H_{n-2}.$$

(39) is a special, *recursive* relationship. An algorithm is called recursive if it prescribes how to calculate any term of a sequence from some preceding terms. Moreover, (39) is linear as well since it gives the rule for calculation in the form of multiplication by *fixed* coefficients and addition. The following sequence, for example, results from a nonlinear recursion:

(40) $$2, 4, 8, 32, 256, 8192, \ldots .$$

Here every term is equal to the product of the two preceding terms. The n-th term of (40) is 2^{F_n} where F_n is the n-th element of the Fibonacci sequence.

Recursive algorithms have special importance because of the fact that a number-sequence which can be calculated by a recursive algorithm can be programmed very easily. The computer has only to repeat the same simple program using different numbers. Such programs are called *cyclic* programs.

FINAL CADENCE

These "variations" on the Fibonacci numbers have led us to many interesting problems in algebra, number theory, combinatorics, geometry, difference and differential equations, search theory, recursive algorithms and the Monte-Carlo method. The sequence of variations could be continued but we think this has been enough to get across the idea that a simple mathematical problem, pursued persistently, can offer a view of many topics in modern mathematics, just as a simple tune frequently has more to it than one would have thought at the first hearing.

The mathematical theory of trees*

INTRODUCTION

The aim of this lecture is to introduce you to the mathematical theory of trees and to give an overview of its many possible applications (in algebra, information theory, operation research, chemistry, biology, etc.).

In the title, I have used the adjective "mathematical" to avoid any misunderstanding, although it might have been even better to use the title "The theory of mathematical trees", since the "trees" about which I am going to speak are mathematical objects, graphs of a certain type. The theory of trees is a chapter of graph theory.

V. W. Rouse-Ball in his famous book "Mathematical recreations and essays" [1] (first published in 1892) devoted a whole chapter to this subject (see [1], pp. 260–262) and called them "geometrical trees". Considering what is meant in the present day by the term "geometry", graphs and especially trees do not belong to the field of geometry, notwithstanding the planar geometrical representation of graphs (with figures consisting of points ("vertices") and arcs ("edges") connecting certain pairs of points). In spite of these, a graph is not a geometrical object since the whereabouts of the points on the plane is unimportant; what counts is which pair of points are connected. Therefore the three drawings in Fig. 1 represent the same graph.

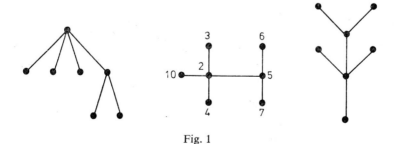

Fig. 1

---

* This is the text of the "Rouse-Ball Lecture" given in Cambridge on April 30, 1968.

At one time, graph theory was considered a branch of topology, but are connected by a path. A connected graph is said to be a tree if it contains no cycle.

Obviously, any two vertices of a tree are connected by one and only one path. The length of the longest path in a graph is called the "diameter". A graph with no cycles can still be unconnected; in such a case, it consists of connected parts called "components" which are all, of course, trees. A graph with no cycles is said to be a "forest". Every connected graph having n vertices contains trees of n-vertices called spanning trees of the graph. If the graph is not a tree, then it contains at least three such trees.

The nomenclature "tree" was introduced into the literature by A. Cayley in one of his first papers [2] in 1857. There followed three other papers [3], [4], [5]. The first important findings concerning trees were given by Cayley, the development of work on the theory of trees begun at Cambridge University. However, the notion of the tree is older. For example, ten years before Cayley, in 1847, Kirchhoff had investigated trees in connection with the study of electrical networks.

Family trees were common in the 17th century, and probably even before. Looking even farther back, the first mathematical book published in Europe in the Middle Ages, Fibonacci's work of the 13th century contained a problem concerning the multiplication of rabbits which led to the famous sequence 1, 1, 2, 3, 5, 8, ... subsequently named after the author. This problem can be interpreted as a question of counting the terminal nodes of certain trees. (See Fig. 2.)

We will return to this problem when investigating the connection between trees and branching processes.

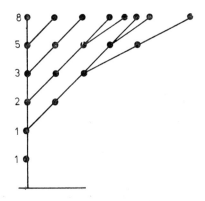

Fig. 2

ENUMERATING PROBLEMS RELATING TO TREES

The number of the vertices of a tree is called its order. The simplest theorem on trees is the following: the number of edges (N) of a tree with n vertices is $n-1$. (See Fig. 3 where trees with 1, 2, 3, 4 and 5 vertices are shown.)

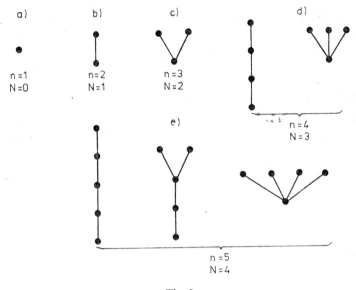

Fig. 3

The proof can be given quite briefly. Choose an arbitrary vertex of the tree and call it the tree's root. Let the numbers $1, 2, ..., n-1$ denote the other vertices. Assign the same number to every edge as denotes the end point of the path starting from the root and ending with the particular edge. Thus every edge is enumerated and every one of the numbers $1, 2, ..., n-1$ has been used exactly once (since every vertex is connected to the root by exactly one path). Hence the number of edges is $n-1$.

The first non-trivial question concerning trees (phrased and answered by Cayley) is this: how many different trees can exist with a given number of vertices? First of all, when are two trees considered to be different? There can be two meaningful answers to this question, one when the vertices are assumed to be distinguishable, by assigning numbers to the vertices of the trees, and the other when the vertices are indistinguishable. Using either assumption, there is no difference for $n=1$ and $n=2$ in the number of distinct trees: there is only one tree of one vertex and one of two vertices (see Figure 3a and 3b). If the vertices are numbered, then there are three different trees with three vertices, while they are all the same if the vertices are not labelled.

Let us denote the number of trees with n labelled vertices as C_n (for brevity we will call these labelled trees). Cayley proved that

$$C_n = n^{n-2},$$

that is

$$C_1 = 1^{-1} = 1, \quad C_2 = 2^0 = 1, \quad C_3 = 3^1 = 3, \quad C_4 = 4^2 = 16, \quad \text{etc.}$$

The labelled tree of 16 vertices can be derived from the two types of trees shown in Fig. 3d as follows: the first type (having one path of length 3) can be numbered in 12 ways (each of the 24 permutations gives the same tree as the reversed permutation); the second type can be labelled in 4 ways depending on the numbering of the center. Prüfer [6] gave the most elegant proof of the Cayley theorem. The main idea behind this proof is to use a sequence of $n-2$ numbers each term of which is one of the numbers $1, 2, ..., n$. One such sequence is called the Prüfer code word of the tree. Since there are n^{n-2} such code words, it is enough to show that there exists a bijective correspondence between the numbered trees with n vertices and the Prüfer code words. This fact can be demonstrated in the way that the coding/decoding is carried out. Before I give the rules for coding and decoding, two simple notions should be introduced. The degree (valency) of a vertex P of a graph G is the number of edges in G one of whose endpoints is P. A vertex P of degree 1 is said to be a "terminal vertex" of the graph. An edge with at least one endpoint being terminal is called a terminal edge. A tree of at least 2 vertices has at least 2 terminal vertices. Now we can turn to the proof of Cayley's theorem given by Prüfer. The coding algorithm is as follows:

a) Let us take the terminal vertex labelled with the smallest number (no. 3), remove it from the tree along with the terminal edge leading to it and write down the index of the other endpoint of this edge (no. 4) as the first digit of the code word.

b) Repeat a) with the remainder of the tree until there remains a graph of only 2 vertices, then stop. Fig. 4 shows an example.

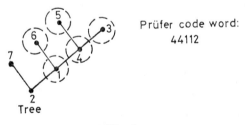

Fig. 4

It is easy to see that this procedure is uniquely reversible; thus the tree can be reconstructed from the code word.

The algorithm for decoding is as follows: write under the code word in increasing order those numbers from $1, 2, ..., n$ which do not appear in the code word. Let us call the resulting sequence an anticode. Connect the vertex numbered with the first digit in the code word to the vertex labelled with the first term of the anticode word. Drop these digits, if the deleted part of the word is not repeated anywhere else in the code word, write it in the anticode at the appropriate place (i.e., preserving the increasing order of this sequence). Repeat the procedure with the new code word and anticode until the code word disappears. Finally, connect the two points having as indices the two remaining digits in the anticode.

For an example, see Fig. 5.

Prüfer code word: 6233

Procedure for decoding:

$$\begin{array}{ccccc} 6233 & 233 & 33 & 3 & \\ 145 & 456 & 256 & 56 & 36. \end{array}$$

The corresponding tree is:

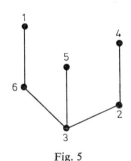

Fig. 5

From this construction, it can be seen that if vertex k of the tree has degree d, then k will appear in the Prüfer code word of the tree exactly $(d-1)$ times; accordingly, only indices belonging to terminal vertices will not be in the code word.

A complete graph with n vertices, has n^{n-2} labelled spanning trees.

A tree is said to be rooted if it has a distinguished vertex: its root. It follows from the Cayley theorem that the number of rooted trees with n labelled vertices is n^{n-1}.

The enumeration of unlabelled trees is more difficult. Cayley solved this problem as well. The nine possibilities for an unlabelled tree of 5 vertices are shown in Fig. 6:

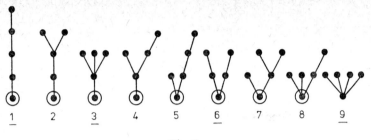

Fig. 6

The above-mentioned result arrived at by Cayley was applied by Rouse-Ball (see [1]) to the following question: how many ways can n pairwise non-intersecting circles be arranged?

For example, in the case of $n=3$, there are four possibilities which are shown in Fig. 7.

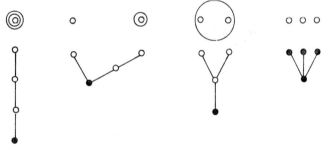

Fig. 7

Rouse-Ball showed that the number of possibilities for n circles is equal to the number of unlabelled trees with $n+1$ vertices (i.e., for $n=3$, this number will be 4). This fact can be seen from the following construction. Take any one of the possible configurations of n circles. Include this configuration in a big circle and consider each of the $n+1$ circles to be a vertex of a graph. Connect two vertices if one of the circles is contained in the other without a third between them. The root is the circle containing the whole configuration. The four possible arrangements of three circles and their corresponding trees can be seen in Fig. 7.

As Rouse-Ball showed, it follows from Cayley's result that if we number n circles, the number of possible configurations is $(n+1)^{n-1}$. For example, if $n=3$ (see Fig. 7), the 4 configurations can be labelled in a total of 6, 6, 3 and 1 ways, and

$$6+6+3+1 = 16 = (3+1)^{3-1}.$$

STATISTICAL THEORY OF TREES

How many terminal vertices can a tree with n vertices have? At first glance, the answer to this question would seem to be trivial. This number t is always ≥ 2 and $\leq n-1$, and these extreme cases (a path or a star) are just as possible as any of those between. However, the question is not yet answered thus: in fact, if n is very large, the cases when $t=2$ or $t=n-1$ will be relatively rare among the very many possibilities (in case of numbered trees as we have already seen there are n^{n-2}). So, $t=n-1$ in n cases while $t=2$ in $\frac{n!}{2}$ cases, and these numbers are quite small compared to n^{n-2} (as can be shown by means of the Stirling formula). Let us now modify the question: how many numbered trees are three with n vertices having exactly t terminal vertices? What is the number of terminal vertices on most of the (labelled) trees with n vertices? This last question can be phrased in another way: if the n^{n-2} distinct (labelled) trees with n vertices are put into a big hat, and if one is taken out at random and its terminal vertices are counted, what can we say about the approximate possible value of this number? Or: approximately how many terminal vertices does a typical tree of n vertices have?

A few years ago, I published [7] the solution of this problem. The answer to the question set out above is that the number of terminal vertices of most trees is approximately $\frac{n}{e}$, where e is the base of the natural logarithm ($e=2.7182...$); that is, 36.8% of the vertices of a typical tree are terminal vertices. Similarly we can ask: how many vertices of degrees d does a typical tree with n vertices have? The answer is: approximately $\frac{n}{e(d-1)!}$; that is, there are approximately $\frac{n}{e}$ vertices of degree 2, $\frac{n}{2e}$ for degree 3, $\frac{n}{6e}$ for degree 4, etc. Note that $\sum_{d=1}^{\infty} \frac{n}{e(d-1)!} = 1$.

This means that the degree minus 1 of the vertices of a typical tree follows a Poisson distribution with a mean equal to 1. About the average degree: it is easy to see that this is always $2-\frac{2}{n}$. Indeed, one can easily see that the sum of the degrees of the vertices in a graph is always equal to twice the number of its edges, i.e., for a tree with n vertices, it is $2n-2$.

The proof of this fact follows from the Prüfer coding of trees. This coding can be interpreted as follows: the choice at random of one of all the possible labelled trees with n vertices (assuming that every one of the n^{n-2} is equally probable) can be carried out by placing $n-2$ balls into n boxes (or urns) randomly in such a way that every ball can end up in any box with the same probability. The number of boxes remaining empty equals the number of terminal vertices, the number of boxes containing exactly one ball equals the number of vertices of degree 2 and, in general, the number of boxes containing exactly $d-1$ balls corresponds to the number of vertices of degree d in a randomly chosen tree. Phrased in this way, the problem can be solved easily. It can be shown that the distribution of trees according to the number of their terminal vertices is approximately normal with an expected value of $\sim \dfrac{n}{e}$ and a standard deviation of $\dfrac{n}{e}\left(1-\dfrac{2}{e}\right)$.

The distribution of trees on the basis of the number of their vertices having degree d is also approximately normal*, with an expected value of $\dfrac{n}{e(d-1)!}$ and a standard deviation of $\dfrac{n}{3(d-1)!}\left(1-\dfrac{1+(d-2)^2}{e(d-1)!}\right)$ (see Fig. 8).

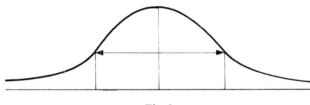

Fig. 8

Let us now investigate, from a statistical point of view, the problem of the height h of a rooted tree, that is, the length of the longest path starting from the root. (The length of a path is the number of edges making up the path). We recently answered this question with György Szekeres (see [9]). The corresponding non-statistical problem can once again be answered easily: the height is at least 1 and at most $n-1$; these and all values in between are possible.

We found that the height of a typical rooted tree with n vertices asymptotically approaches the value $\sqrt{2n\pi} \approx 2.50663\sqrt{n}$. We also determined the distribution of trees on the basis of their height. It can be shown that the diameter

---

* This result can be proven using the solution to the classical urn problem given by J. V. Bolotnikov (see [8]).

of a typical tree with n vertices is of the order of $\sqrt{n}$. The limit distribution of $\dfrac{P}{\sqrt{n}}$ is not known!

We should mention that Cayley also investigated the problem of how many rooted trees there are of height h with n vertices.

APPLICATIONS IN OPERATION RESEARCH

Let us take the following practical problem: a network (cable, highways, railways, etc.) has to be built in such a way that it will connect any two of a certain number of cities. The total cost of the network must be minimized. Moreover, the cost of directly connecting any two cities is given. Obviously, the optimal network is necessarily a tree. As we have seen there are n^{n-2} trees connecting n cities so the optimal network should be one of them. An algorithm for solving this problem leading step-by-step to the optimal solution (Fig. 9) was given by Boruvka [10] (see also Kruskal [11]).

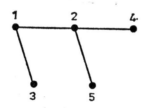

Fig. 9

Costs:

$C(1, 2) = 1$

$C(1, 3) = 1.5$

$C(1, 4) = 3$

$C(1, 5) = 6$

$C(2, 3) = 2$

$C(2, 4) = 1.5$

$C(2, 5) = 1.2$

$C(3, 4) = 8$

$C(3, 5) = 4$

$C(4, 5) = 3$

Another problem of operation research leading to trees under certain conditions is the following. Assume that we wish to establish air-routes between n given airports so that one can travel from any one airport to any other changing planes at most $d-1$ times. Another restriction is that the maximal capacity of the airports is fixed, so that the number of planes leading and arriving at a given airport cannot exceed the number k. In the language of graph theory: construct a connected graph with n vertices, of diameter $\leq d$, where the degree of the vertices cannot be greater than k and the number of edges is minimal. Pál Erdős and Vera T. Sós have shown in their paper [12] that if k is not smaller than a certain lower bound $U_d(n)$, then the optimal graph is always a tree (having $n-1$ edges). For example, if $d=3$ (i.e., if one would have to change planes at most twice starting from any airport in the network to travel to any other) this lower bound is $\dfrac{n}{2}$. The optimal networks of diameter 3 for $n=10$ and $k=3, 4$ and 5 are shown in Fig. 10. We can see that for $k=5=\dfrac{10}{2}$, the optimal graph is a tree.

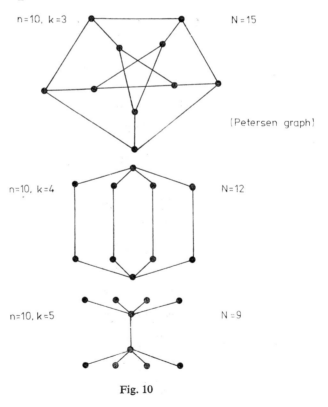

Fig. 10

TREES AND INFORMATION THEORY

Cayley made the following observation: given a binary operation $A \circ B$, which is not associative, that is, $(A \circ B) \circ C$ is not necessarily equal to $A \circ (B \circ C)$, as $(A^B)^C \neq A^{(B^C)}$), then the symbols $A \circ B \circ C$, $A \circ B \circ C \circ D$, etc., have more than one meaning which will be specified only if the order of the operations is made clear by the use of appropriate brackets. For example, $2^{2^{3^2}}$ can be interpreted in five ways:

$$2^{[2^{(3^2)}]} = 2^{512}; \quad [(2^2)^3]^2 = 2^{12}; \quad 2^{[(2^3)^2]} = 2^{64};$$

$$(2^2)^{(3^2)} = 2^{18}; \quad [2^{(2^3)}]^2 = 2^{16}.$$

Cayley also asked the following question: how many meanings can an expression $A_1 \circ A_2 \circ \ldots \circ A_n$ have? What he found was that to every interpretation of the above symbol sequence there corresponds a rooted tree of n terminal vertices with all of its non-terminal vertices (excluding the root) being of degree 3. Such a tree is said to be a binary-code tree, because every terminal vertex can be coded with a sequence of zeroes and ones. There is only one path to every terminal vertex; following this path, we write a 0 for each left turn and a 1 for each right turn. Such trees are frequently called binary trees.

Figure 11 shows the trees corresponding to the different interpretations of the expression $A \circ B \circ C \circ D$: Note that in a binary-code tree with n terminal vertices there are $n-1$ further vertices including the root. The following question can now be asked: how many binary-code trees of n labelled terminal vertices are there? Rouse-Ball found this number (B_n) to be:

$$B_n = \frac{1 \cdot 3 \cdot 5 \cdot \ldots \cdot (2n-3) 2^{n-1}}{n!} = \frac{1}{2n-1} \binom{2n-1}{n}.$$

Binary-code trees play an important role in information theory. Let us assume that we wish to code a certain number of messages with sequences of different lengths consisting of zeroes and ones in such a way that the code is "without a comma", that is, if we write the code words one after another without separating them by any kind of symbol, we can still determine where one code word ends and another begins. This condition is satisfied through the use of a *prefix* code, where no code word is the continuation of any other (but if any digit is omitted from any of the code words, the prefix property of the code will no longer hold). To every such code there corresponds a rooted binary tree with as many terminal vertices as code words, and vice versa. For

Enumerating principle:

((AoB)oC)oD

    0 0 0
    0 0 1
    0 1
    1

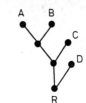

Ao(Bo(CoD))

    0
    1 0
    1 1 0
    1 1 0

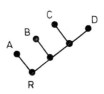

(AoB)o(CoD)

    1 0
    1 1
    0 0
    0 1

(Ao(BoC))oD

    0 0
    0 1 0
    0 1 1
    1

Ao((BoC)oD)

    0
    1 0 0
    1 0 1
    1 1

A : 0   C : 1 0 1
B : 1 0 0   D : 1 1

Fig. 11

example, for the code

$$111$$
$$1101$$
$$1100$$
$$10$$
$$01$$
$$001$$
$$000$$

the corresponding code tree is shown in Fig. 12:

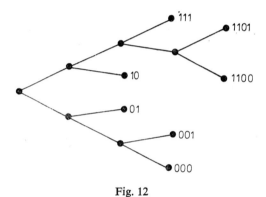

Fig. 12

If the probabilities of the code words are given, then the best code is the one which has a minimal average length for the given distribution. Such an optimal code, or the construction of the corresponding tree, can be arrived at with the assistance of Huffman's algorithm.

Binary code trees can also be interpreted using the terminology of search theory. Under such circumstances, every vertex corresponds to a question answerable with a "yes" or a "no", and to each of these answers there belongs an edge emanating from the particular vertex.

For example, if someone would like to interview different people so that every question asked depends on the answer given to the previous question, then the plan of the interview can be depicted as a binary tree.

TREES AND PERMUTATIONS

In the library of the Mathematical Institute of Oberwolfach, visitors are requested to return books to the appropriate places by a witty sign which runs as follows: "Do not forget that the transposition of neighbouring elements generates the complete symmetric group!" This means that any permutation of the numbers $1, 2, ..., n$ can be obtained by successive applications of operations each of which consists of the transposition of a number k with $k+1$. For example, starting with the increasing sequence 1234, we can arrive at the permutation 4321 as follows:

in 1234 transpose 3 and 4 to get 1243
in 1243 transpose 2 and 3 to get 1342
in 1342 transpose 1 and 2 to get 2341
in 2341 transpose 3 and 4 to get 2431
in 2431 transpose 2 and 3 to get 3421
in 3421 transpose 3 and 4 to get 4321

In general, we can ask when $(n-1)$ given transposition (i, j) will generate the complete symmetric group. The answer: if and only if the edges (P_i, P_j) are the edges of a tree consisting of the vertices $P_1, P_2, ..., P_n$; this was proven by György Pólya [13].

12<u>34</u>
12<u>43</u>
1<u>34</u>2
2<u>3</u>41
2<u>43</u>1
<u>34</u>21
4321

An arbitrary set of transpositions will generate the complete symmetric group if the corresponding graph (called a Pólya-graph) is connected, that is, if it contains a tree with n vertices (see [14]).

For example, the transpositions $(1, 2), (2, 3), (2, 4)$ are associated with a tree, and the permutation 4321 can be derived from these transpositions as follows:

1<u>2</u>34
1<u>32</u>4

(4321) = (23)(12)(24)(12)   231<u>4</u>
                               43<u>12</u>
                               4321

From the fact that n^{n-2} trees can be composed of n vertices, it follows that there are n^{n-2} distinct systems, each of which consists of $n-1$ transpositions and generates the complete symmetric group of n elements.

TREES AND CHEMISTRY

Cayley also investigated the probable structure of paraffin whose molecules can be described with the formula C_nH_{2n+2}. Every carbon atoms has valence 4 and every hydrogen atom's is 1.

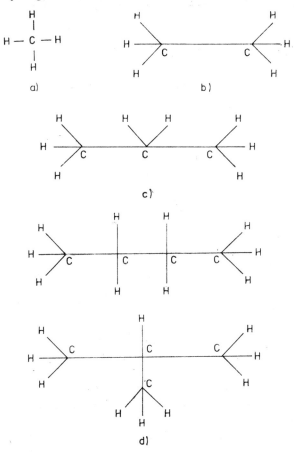

Fig. 13

In particular can be seen in Fig. 13a, b, c. the structure of methane (CH_4), of ethane (C_2H_6) and of propane (C_3H_8) are as follows:

The structure of butane is ambiguous, since there are two possible isomers (see Fig. 13d).

The paraffin molecule is a tree. If we remove the hydrogen atoms, the graph of the carbon atoms which remains will still be a tree (since every H is a terminal vertex) where every vertex has a degree of at most 4.

Therefore, the number of possible structures of paraffin molecules, C_nH_{2n+2}, is equal to the number of labelled trees each of whose vertices has a degree ≤ 4. Thus, this question too, has led us to the enumeration problem related to certain trees.

These investigations were carried further by Pólya and others.

TREES AND BIOLOGY

Earlier, we mentioned genealogical trees. They play an important role in the theory of branch processes. For simplicity, I will treat only the reproduction of bacteria. Let us assume that, after certain time span, every bacterium either divides into two or ceases to exist. Under such conditions, the genealogical tree of a bacterium is a binary code tree.

I will mention only the following question here: in how many ways is it possible for a bacterium to have an n-th generation of exactly $2k$ members. Let us denote the number of possibility by $T(n, 2k)$. It can easily be shown that the following recursive formula holds:

$$T(n+1, 2k) = \sum_{l \geq \frac{k}{2}} T(n, 2l) \binom{2l}{k}$$

so that the appropriate generating function*

$$P_n(x) = \sum_k T(n, 2k) x^{2k}$$

---

* In general, if $T(k)$ is a function defined on the positive integers (usually the solution for $k=0, 1, 2, \ldots$ of a combinatorial problem) then by its generating function we mean the infinite sum

$$P(x) = T(0) + T(1)x + T(2)x^2 + \ldots .$$

If $T(k)=0$ is beyond a certain k, then the sum will be finite. (Gy. Katona)

will satisfy the recursive formula:

(+) $$P_{n+1}(x) = P_n(1+x^2),$$

so that
$$P_1(x) = 1+x^2$$
$$P_2(x) = 2+2x^2+x^4,$$
$$P_3(x) = 5+8x^2+8x^4+4x^6+x^8,$$

and so on. For an example see Fig. 14.

$$T(3, 4) = 8.$$

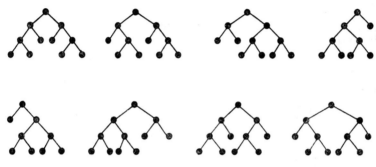

Fig. 14

The recursion under (+) is a special case of a more general formula for the Galton–Watson process in the theory of branching processes. Although there is more that could be said about the theory of trees and its applications, I hope that what I have said provides a glimpse into the characteristics of the theory and its wide applicability in various branches of mathematics.

BIBLIOGRAPHY

[1] W. W. Rouse-Ball: *Mathematical Recreations and Essays*. 11th ed., revised by H. S. M. Coxeter, MacMillan Co., London 1947.
[2] A. Cayley: On the theory of the analytical forms called trees. I., *Phil. Mag.* **13**, 1857. 172–176, Coll. Papers 203.
[3] A. Cayley: On the theory of the analytical forms called trees. II., *Phil. Mag.* **18**, 1859. 374—378, Coll. Papers.
[4] A. Cayley: A theorem on trees. *Quarterly J. of Pure and Appl. Math.* **23**, 1889. 376–378, No. 895.
[5] A. Cayley: On the analytical forms called trees with applications to the theory of chemical combinations. *Rep. British Assoc. for the Advancement of Sci.* 1875. 257–305.

[6] A. Prüfer: Neuer Beweis eines Satzes über Permutationen. *Arch. für Math. u. Phys.* **27,** 1918. 142–144.

[7] A. Rényi: On the theory of trees. *Publ. Math. Inst. Hung. Acad. Sci.* **4,** 1959. 73–85.

[8] Yu. V. Bolotnikov: Convergence of the variables $\mu-(n)$ to Gaussian and Poisson processes in the classical problem with balls. *Teoria Veroiatnostei* **13,** 1968, 39–50.

[9] A. Rényi, G. Szekeres: On the height of trees. *J. Australian Math. Soc.* **7,** 1967. 497–507.

[10] O. Boruvka: On a minimal problem. *Prace Moravske Pridovedecke Spolecnosti,* **3,** 1926.

[11] W. Kruskal: The shortest spanning subtree of a graph and the traveling salesman problem. *Proc. Amer. Math. Soc.* **7,** 1956, 48–50.

[12] P. Erdős, A. Rényi, Vera T. Sós: On a problem of graph theory. *Studia Sci. Math. Hung.* **1,** 1966, 215–235.

[13] G. Pólya: Kombinatorische Anzahlbestimmungen für Gruppen, Graphen und chemische Verbindungen. *Acta Math.* **68,** 1937, 145–255 (especially 208–209).

[14] P. Erdős, A. Rényi: On the evolution of random graphs. *Publ. Math. Inst. Hung. Acad. Sci.* **5,** 1960. 17–61.

Index

Abel, N. H. 30
Abelian group 59
automation 5

Bar-kochba 13, 20
—, with lies 47
binary code 36
—, tree 115, 117
binary digit 7, 36
binary number 6
binary primitive prefix code 37
binary representation 36
binary system 7
Binet-formula 96
bit 7, 11, 19, 20, 22
blackjack 74
Bolotnikov, J. V. 1
Boltzmann 18, 19
Boltzmann—Shannon formula 18
Borel, E. 66
Borel- set 72
Boruvka, O. 113
bridge 63

card, well shuffled 57
Cayley, A. 106—110, 115, 119
Cayley theorem 108—109
channel 43, 53
channel capacity 5 45
Charon, A. 66
codename 8
code tree 37
code word 36, 109
—, Prüfer 108—109, 111
code word length, avarage 39, 53
—, expected 39

coding 5, 7, 108
—, error correcting 40
—, optimal 117
coding theorems 45
component 106
computer 6
—, to translate 41
conditional distribution 23
conditional entropy 23
conditional probability 22—23, 27
Crelle 30
Culbertso system 63, 65
cyclic group 59
cyclic program 104

decoding 36, 43—44, 53, 108
dependence 25
deterministic model 58
diameter 106
digit 7, 8, 36
distribution, conditional 23
distribution of cards 60
distribution of trees 112
Dubbins, L. E. 73

edge, terminal 108
element 9
emitter 43
encoder 43
entropy 18, 19, 23, 26, 52, 53
error- correcting code 40
event, observable 83
—, unexpectedness of 26
—, unobservable 83
experiment 83

feedback 5, 44
Fibonacci, L. 86, 106
Fibonacci numbers 89, 90, 92, 99—103
Fibonacci sequence 85, 86, 88—90, 99, 100—103
forest 106

Galton-Watson process 121
game of chance 56, 66
generating function 120
geometrical tree 100
Golden section 97, 99
group 57
Hartley 10
Hartley's formula 9, 10, 12, 14, 17
haphazard element 12
Huffman code 43, 54, 117

indicator of event 28
information 1, 22
—, amount 7, 9, 19
—, bit of 11, 19, 20, 22
—, concept 6
—, conservation 39
—, flow 5
—, full 24
—, law of additivity 9, 19, 34
—, law of conservation 23
—, measure of 6, 7
—, mutual 34
—, relative 24, 27
—, speed of transmission 41, 45
—, unit of 6, 7
information current density 52
information theory 1, 4, 5, 115

Jordan, K. 63

Kiefer, I. 99
Kirchoff 106
Kruskal, W. 113

large numbers, law of 16
Lebesque-measure 73

Markov chain 59
martingale 67

mathematics in physics 32
measure 72
message 9
Morse code 54
mutual information 34

noise 43
noiseless channel 53
noisy channel 43, 44

operation 59

Pascal triangle 100, 101
path 106
periodic sequence 101
permutation 118
physics teaching 32
Poisson distribution 111
Pólya, Gy. 118, 120
Pólya-graph 118
prefix code 36, 115
—, primitive 37, 41
probability, conditional 22, 23, 27
probability theory, teaching 77—84
probability space 83
Prüfer 108
Prüfer code word 108, 109, 111
pseudo-random numbers 103

receiver 43
recursive algorithm 104
redundancy 39
Révész, P. 82
Rham, G. de 72
root 107, 109
roulette 66
Rouse-Ball, V. W. 105, 110, 115

Sarage, L. J. 73
search theory 117
sequence 6, 8, 101
set 9
Shannon 18
Shannon formula 14, 16, 17, 18, 20, 22, 27
shuffling model 59
signal 51

INDEX

Sós, V. T. 114
spanning trees 106
Stirling formula 111
stohastic model 58
symmetric group 118, 119
Szekeres, Gy. 112

terminal edge 108
terminal vertex 108
theory of tree 105
Thorp, E. O. 74
transmission, 7, 44
tree 37, 105—109, 115, 119
—, bynary- code 115

tree, geometrial 105
—, labelled 108
—, spanning 106
—, unlabelled 109
Turán, P. 59
"Twenty Qusetions" 5
uncertainty 28

unlabelled tree 109
unexpectedness 26, 28

vertex 108
Wiener 18
winning strategy 67

Applied Probability and Statistics (Continued)

JUDGE, HILL, GRIFFITHS, LÜTKEPOHL and LEE • Introduction to the Theory and Practice of Econometrics

JUDGE, GRIFFITHS, HILL, LÜTKEPOHL and LEE • The Theory and Practice of Econometrics, *Second Edition*

KALBFLEISCH and PRENTICE • The Statistical Analysis of Failure Time Data

KISH • Survey Sampling

KUH, NEESE, and HOLLINGER • Structural Sensitivity in Econometric Models

KEENEY and RAIFFA • Decisions with Multiple Objectives

LAWLESS • Statistical Models and Methods for Lifetime Data

LEAMER • Specification Searches: Ad Hoc Inference with Nonexperimental Data

LEBART, MORINEAU, and WARWICK • Multivariate Descriptive Statistical Analysis: Correspondence Analysis and Related Techniques for Large Matrices

LINHART and ZUCCHINI • Model Selection

LITTLE and RUBIN • Statistical Analysis with Missing Data

McNEIL • Interactive Data Analysis

MAINDONALD • Statistical Computation

MANN, SCHAFER and SINGPURWALLA • Methods for Statistical Analysis of Reliability and Life Data

MARTZ and WALLER • Bayesian Reliability Analysis

MIKÉ and STANLEY • Statistics in Medical Research: Methods and Issues with Applications in Cancer Research

MILLER • Beyond ANOVA, Basics of Applied Statistics

MILLER • Survival Analysis

MILLER, EFRON, BROWN, and MOSES • Biostatistics Casebook

MONTGOMERY and PECK • Introduction to Linear Regression Analysis

NELSON • Applied Life Data Analysis

OSBORNE • Finite Algorithms in Optimization and Data Analysis

OTNES and ENOCHSON • Applied Time Series Analysis: Volume I, Basic Techniques

OTNES and ENOCHSON • Digital Time Series Analysis

PANKRATZ • Forecasting with Univariate Box-Jenkins Models: Concepts and Cases

PIELOU • Interpretation of Ecological Data: A Primer on Classification and Ordination

PLATEK, RAO, SARNDAL and SINGH • Small Area Statistics: An International Symposium

POLLOCK • The Algebra of Econometrics

PRENTER • Splines and Variational Methods

RAO and MITRA • Generalized Inverse of Matrices and Its Applications

RÉNYI • A Diary on Information Theory

RIPLEY • Spatial Statistics

RIPLEY • Stochastic Simulation

RUBIN • Multiple Imputation for Nonresponse in Surveys

RUBINSTEIN • Monte Carlo Optimization, Simulation, and Sensitivity of Queueing Networks

SCHUSS • Theory and Applications of Stochastic Differential Equations

SEAL • Survival Probabilities: The Goal of Risk Theory

SEARLE • Linear Models

SEARLE • Matrix Algebra Useful for Statistics

SPRINGER • The Algebra of Random Variables

STEUER • Multiple Criteria Optimization

STOYAN • Comparison Methods for Queues and Other Stochastic Models

TIJMS • Stochastic Modeling and Analysis: A Computational Approach

TITTERINGTON, SMITH, and MAKOV • Statistical Analysis of Finite Mixture Distributions

UPTON • The Analysis of Cross-Tabulated Data

UPTON and FINGLETON • Spatial Data Analysis by Example, Volume I: Point Pattern and Quantitative Data

(continued from front)